TRANSACTIONS

OF THE

AMERICAN PHILOSOPHICAL SOCIETY

HELD AT PHILADELPHIA

FOR PROMOTING USEFUL KNOWLEDGE

VOLUME XXI—NEW SERIES

PART III

ARTICLE III—*Chromosomes in the Spermatogenesis of the Hemiptera Heteroptera.* By *Thos. H. Montgomery, Jr.*

Philadelphia:

THE AMERICAN PHILOSOPHICAL SOCIETY

104 South Fifth Street

1906

ARTICLE III.

CHROMOSOMES IN THE SPERMATOGENESIS OF THE HEMIPTERA HETEROPTERA.*

By Thos. H. Montgomery, Jr.

The present paper treats of the behavior of the chromosomes in forty species of the Hemiptera, whereby especial attention is given to their number and form in the maturation mitoses, and to the changes of the modified chromosomes. Then there are treated from broader points of view, the modified chromosomes, chromosome difference, and the facts of the number of chromosomes. This is an amplification and correction of earlier researches of mine (1898, 1901a, 1901b, 1904a) upon the same species; and the preparations studied were the same as those previously used.

Certain phenomena treated in those earlier papers are not discussed in the present one, such as the conditions of the plasmosomes (nucleoli), and the relations of the modified chromosomes in the rest stage of the spermatogonium.

I have felt it necessary to introduce a new nomenclature, indicated in a preliminary note (1906), for the different kinds of chromosomes. Since the discovery of peculiarly modified chromosomes in certain of the insects a great variety of names has been proposed for them, and most of these suffer from a quite unnecessary length. My own earlier terms "heterochromosome" and "chromatin nucleolus" were cumbersome, and "accessory chromosome" and "heterotropic chromosome" sin equally in this regard, while "special chromosome" and "idiochromosome" are no way self-explanatory. Therefore for the sake of uniformity but more especially simplicity in writing I here employ the following nomenclature:

Chromosome, the original term of Waldeyer (1888), to be retained as a convenient collective word for each separate mass of chromatin and linin. When there are no marked differences in the behavior of the several chromosomes of a cell, all may be given this name. But when chromosomes of different behavior occur, they are distinguished as follows:

(1) *Autosome (autosoma)*, the non-aberrant chromosomes that I have previously called *ordinary chromosomes*.

(2) *Allosome (allosoma)*, any chromosome that behaves differently from the autosomes, and is a modification of the latter. This term is much more concise than my

* Contributions from the Zoological Laboratory of the University of Texas, no. 72.

A. P. S.—XXI. J. 21, 7, '06.

earlier one, *heterochromosome*, and etymologically has the same significance. Two main kinds of allosomes are now known in spermatogenetic cycles, and these are :

(*a*) *Monosome* (*monosoma*), an allosome that is unpaired in the spermatogonium, *i. e.*, without a correspondent mate there. Heretofore these have been named variously : *accessory chromosomes* (McClung), *chromosomes spéciaux* (de Sinéty), *chromosomes x* and *unpaired chromosomes* (Montgomery), *heterotropic* and *differential chromosomes* (Wilson).

(*b*) *Diplosome* (*diplosoma*), allosomes that occur in pairs in the spermatogonium. These have been previously denominated : *small chromosomes* (Paulmier), *chromatin nucleoli* (Montgomery), *idiochromosomes* and *m-chromosomes* (Wilson).

I regret to have to add new names to the cytological dictionary, for there is already somewhat of a chaos of them. But these seem to be about as simple and uniform as could be invented, and I trust that their convenient brevity will insure their adoption by fellow investigators.

Wilson's recent series of " Studies on Chromosomes " has brought out two new and important points with regard to the allosomes. One is that the diplosomes (his idiochromosomes) of certain Hemiptera conjugate in the second spermatocytes and there divide reductionally. This phenomenon had been entirely overlooked by me ; my oversight was due in part to the fact that in most of the species I did not examine the spermatogenesis beyond the stages of the first maturation mitosis ; and in greater part to the fact that I was influenced by the thought that when there is an even number of chromosomes in the spermatogonium there must be exactly half that number of bivalent chromosomes in the first spermatocytes. And yet in certain species (*Euschistus tristigmus*, *Oncopeltus*, *Zaitha*), I showed that diplosomes may be univalent in the first spermatocytes and divide there separately. Now I am able to confirm ·Wilson's discovery for quite a number of species. His second and more valuable conclusion is that when there is a single monosome in the spermatogenesis, it is always represented by a pair in the ovogenesis ; and Miss Stevens and he have enlarged upon this phenomenon to partially explain sex-determination. Further, Wilson has found the occurrence of a monosome in certain Coreids where I had overlooked it, and even in *Anasa* where his own student, Paulmier, had not found it.

The present paper then is an attempt to reconcile these differences of observation, on the basis of a fuller and more complete study of all of my old material. It seemed clearest to present the facts gained for each species separately, then in conclusion to bring them together under certain generalizations.

The term "reduction division " is here used to express the separation of entire chromosomes from each other in an anaphase of division ; ·or, in the case of a monosome, of its passage without division to one of the daughter cells.   In reality such processes are not acts of division at all, but rather ones of separation, yet it seems best to retain the long-accustomed terminology for them.   And by "equational division" is meant any division of a univalent chromosome; this is always along the length of an elongate element, and then probably always an equal halving; in the case of a rounded chromosome it is practically impossible to determine the plane of the division, except by an analysis of the changes of the chromosome in the early prophases, when it can be demonstrated that even rounded chromosomes divide in a plane along which they were previously elongated.

Farmer and Moore (1905) have introduced the term "maiotic phase," "to cover the whole series of nuclear changes included in the two divisions that were designated as heterotype and homotype by Flemming."   But the older word "maturation period" need not be given up, provided we recognize that one of the maturation mitoses is always reductional.

Finally, by the term "safraninophilous" I indicate that an element stains red after the use of the triple stain of Hermann, safranine, gentian violet and orange G; and would again insist on the point that for the study of the allosomes this stain is in a number of ways preferable to the iron hæmatoxyline method.

I.   Observations.

PENTATOMIDÆ.

1.   Euschistus variolarius Pal. Beauv.

Spermatogonic Divisions. — Pole views of the equatorial plate stage show in most cases 14 chromosomes; the two smallest are not quite equal in volume and are the diplosomes (Di, di, Plate IX, Figs. 3, 4); the twelve others are autosomes which compose 6 pairs of graduated volumes (A, a–F, f).   But in one case there were clearly 15, and this was illustrated in Fig. 3 of my preceding paper (1901b); that earlier figure erroneously showed 16 because I had mistaken one of the longest for 2.   And now I find two clear cases each with 16 chromosomes (Figs. 1, 2); the additional elements are the ones marked G, g.   In both of these cells it will be noted that the components of the pair G, g do not lie in the same plane, but that one is placed immediately below the other, which would be a reason to conclude that the two are the precociously separated halves of a single one.   These differences in number are puzzling, and I have been unable to explain them satisfactorily.   But perhaps they are to be interpreted as follows: the usual number of chromosomes is 14, but occasionally there is present an additional one which divides before the others, and thereby gives the appearance of a totality of 16. It was on the basis of cases of this kind that I had previously decided that the normal number is 16, whereas I now find that the usual number is 14. Whenever all the chromosomes lie with their long axes in the plane of the equator their arrangement in pairs of like components may be readily made out.

Growth Period. — In the synapsis the 12 autosomes conjugate to form 6 bivalent ones as I previously described in some detail (1898, 1901*b*). The diplosomes also always unite then end to end. At first each diplosome may become more or less irregularly bent (Fig. 5), later becoming more spherical. After the synapsis period they are at first in intimate contact, each is a little longer than wide with a slight constriction around the middle (Fig. 6, *Di, di*); this probably represents a longitudinal split of each. The two may lie parallel or slightly divergent, or frequently with their long axes making a right angle. When they are so placed a small space is seen between them, and this I erroneously described in 1898 as a vacuole within a single element; now I can decide that no such vacuole is formed, and that the diplosomes swell but little in size during the growth period. Though the two may often be so near together as to appear to form an apparent single sphere, they never seem to actually fuse, for a line of separation can always be found.

First Maturation Division. — The behavior of the autosomes was described in full in the papers already referred to. In the late prophase, just before the dissolution of the nuclear membrane, or at that time, the diplosomes separate. After they separate each may continue to show the longitudinal split (Fig. 8) or may not (Fig. 9); in the latter case there is, that is to say, a temporary closure of the split, just as happens regularly with the autosomes. In the monaster stage are found 8 elements, and all of these are shown on lateral view in Fig. 10. Six of them are bivalent autosomes and these divide reductionally. But each of the two smallest chromosomes is a univalent diplosome, and their division is probably through the plane of their earlier longitudinal split. Each second spermatocyte receives 6 univalent autosomes, and half of each of the diplosomes.

Second Maturation Mitosis. — In the equator of the spindle (Figs. 11, 12) all the 6 autosomes become placed with their constrictions (longitudinal splits) in the plane of the equator, and they all divide equationally. But the two diplosomes conjugate in the middle of the chromosomal plate where they compose a bivalent element with components of unequal volume (*Di, di*), and this double element divides reductionally. Consequently each spermatid receives 7 chromosomes, whereby half the spermatids get the larger diplosome (Fig. 13) and half the smaller (Fig. 14).

Literature. — In my previous papers, 1898, 1901*b*, I made the serious mistake of failing to note the separation of the diplosomes just before the first maturation division, their equational division there, and their conjugation and separation in the second mitosis.   In my first paper on this species the spermatogonial number of chromosomes was correctly given, while in the later paper I was misled by one of the unusual cases, here described, of 16 chromosomes in the equator of the spindle.

2. EUSCHISTUS TRISTIGMUS Say.

Spermatogonic Divisions. — Always 14 chromosomes (Plate IX, Fig. 15), 3 (*Di, F, f*) being noticeably smaller than the others.   When these elements lie suitably 12 of them are seen to compose 6 pairs (*A, a–F, f*) each pair with components of approximately equal volume and form ; these are the maternal and paternal autosomes.   There remain two elements, *Di* and *di*, one of which is the smallest of all, the other larger than this and also larger than either component of the smallest autosome pair ; these two elements of such different volumes are the diplosomes.

Growth Period. — The autosomes unite to form 6 bivalent ones as previously described by me..   The diplosomes also unite regularly and remain so during the earlier part of the growth period (*Di, Di,* Fig. 16), but they later separate.

First Maturation Division. — There are always 8 elements (Figs. 17, 18), 6 of these are bivalent autosomes (*A, a–F, f*), and these divide reductionally.   And 2 are the separated and univalent diplosomes (*Di, di*) which also divide and hence equationally. A pole view of a daughter chromosomal plate of the ensuing anaphase (Fig. 19) before the chromosomes have taken their place in the equator of the second spindle shows the two diplosomes unconstricted, and each of the six autosomes with a constriction that is the longitudinal split.

Second Maturation Division. — In the equator of the spindle (Fig. 20) are seen the 6 autosomes dividing along the line of the longitudinal split ; but the two diplosomes have conjugated end to end and form a bivalent element with unequal components that divides reductionally.   Each spermatid receives 7 chromosomes, half of them receiving the larger (Fig. 22) and half the smaller diplosome (Fig. 21).

In this species each chromosome pair can be followed with great certainty during all its changes, thanks to the marked differences in volume of the different pairs ; and this I have illustrated upon the figures by correspondence in the lettering.

Literature. — My first account was entirely correct (1901*b*), and I described how the diplosomes divide separately in the first maturation mitosis.   But I failed to notice their conjugation in the second spermatocytes.   Wilson's account of this and the preceding species is correct.

3. PODISUS SPINOSUS Dall.

Spermatogonic Divisions. — There are 16 chromosomes in the equator of the spindle (Plate IX, Fig. 23). Fourteen of them make up 7 pairs (A, a–G, g), and the pairs form a gradated series. The 2 others are the diplosomes which are of unequal volumes, one of them (Di) being the smallest of all the chromosomes while the other (di) is as large as the components of the smallest autosome pair.

Growth Period. — The 14 autosomes conjugate to form 7 bivalent ones. The diplosomes likewise become apposed and during the synapsis stage and a part of the later portion of the growth period this bivalent diplosome is placed against the nuclear membrane and is composed of a larger and a smaller element in close contact (Fig. 24, Di, di), but usually, as in the figure, a narrow line of separation is to be seen between the two.

First Maturation Division. — In the late prophases the diplosomes separate, and are apart from each other in the equatorial plate (Fig. 25); the smallest element there is the smaller diplosome (Di), but which element represents the larger it would be difficult to determine from the size relations. Each diplosome divides in the plane of its transverse constriction, which can represent nothing else than a longitudinal split. Each of the 7 bivalent autosomes divides reductionally.

Second Maturation Division. — In the center of the spindle the diplosomes conjugate end to end; Fig. 26 shows a pole view of all the chromosomes, and in the center can be seen a smaller diplosome placed at the end of a larger (Di, di); lateral views (Fig. 27) show clearly this bivalent diplosome with its unequal components. This bivalent element divides reductionally, while all the 7 autosomes divide equationally.

Literature. — My preceding account (1901b) was entirely correct except that I failed to note the unequal volumes of the diplosomes and the phenomenon of their being separate in the first maturation monaster; I had figured and described the second maturation monaster in mistake for the first. Wilson (1905a) was the first to show the conjugation of the diplosomes in the second spermatocyte, and their reductional division there.

4. MORMIDEA LUGENS Fabr.

Spermatogonic Division. — There are apparently 14 chromosomes in the spindle (Plate IX, Fig. 28); this is a redrawing of Fig. 31 of my preceding paper (1901b) in which I had erroneously represented each of the two largest elements A, a as two. There are 6 autosome pairs, A, a–F, f, which show gradations in volumes; only in regard to the supposed pair E, e am I undecided whether it is a single or two chromosomes. The two smallest bodies are the diplosomes (Di, di) and are unequal in size.

Growth Period. — There are formed in the early growth period 6 bivalent autosomes, and one bivalent diplosome.   In the earlier stages the latter is composed of two of unequal volume placed end to end.   Later stages show a much larger, ovoid diplosome containing one large or several smaller vacuoles ; I could not decide whether this is the whole bivalent diplosome or only one of its components.

First Maturation Division. — Pole views of the equatorial plate (Fig. 29) show always 8 elements, 6 of which must be bivalent autosomes.   Two elements are much smaller, and judging by their size relations in the spermatogonia these must be the diplosomes (Di, di) ; if this conclusion be correct, then the bivalent diplosome must have separated into its two elements in the prophases of this mitosis.   The chromosomes are very regularly arranged ; a large autosome forms the center of a circle composed of the five other autosomes and the two diplosomes.

Second Maturation Division. — Pole views show apparently only seven elements in the spindle (Fig. 30) ; but the central one is really bivalent, made up of the two diplosomes placed end to end ; probably this bivalent diplosome undergoes a reduction here, but I cannot say so with certainty because my slides contained only a few of these stages.

Literature. — Previously (1901b) I was mistaken in supposing there to be 16 chromosomes in the spermatogonia ; I did not describe the second maturation division.

5. Cosmopepla carnifex Fabr.

Spermatogonic Divisions. — There are 14 autosomes which compose 7 pairs of gradated sizes (A, a–G, g, Plate IX, Fig. 31) ; and two diplosomes, one of which (Di) is the smallest element of all, while the other is much larger and rod-shaped (di).

Growth Period. — The 14 autosomes conjugate to produce 7 bivalent ones.   The 2 diplosomes also first unite end to end, then more closely side to side ; each of them becomes longitudinally split, and their changes appear to be exactly as described for *Euschistus variolarius.*

First Maturation Division. — In the late prophases (Fig. 32) the diplosomes separate, each is bipartite, and they enter into the spindle apart from each other.   Both of them divide, therefore equationally, while the 7 bivalent autosomes divide reductionally.   On pole views it is difficult to recognize which are the diplosomes (Fig. 33), but on lateral aspects (Fig. 34) they may be recognized as being the two smallest elements and the only ones that are not tetrads.

Second Maturation Division. — Just before the arrangement of the chromosomes in the plane of the equator the unequal diplosomes conjugate in the middle of the equatorial plate to form a bivalent element, hence one sees either 8 bodies (Fig. 35) in which case the smaller diplosome is hidden from view by the larger, or 9 (Fig. 36) when one of the diplosomes is seen below the other. The 7 autosomes divide equationally, but the diplosomes without dividing pass into opposite daughter cells (spermatids). Each spermatid (Fig. 37) shows on pole view 8 chromosomes, a circle of 7 autosomes around a central diplosome; half the spermatids receive the larger diplosome, and half the smaller.

Literature. — I had originally erroneously stated there were 18 chromosomes in the spermatogonia, and had failed to note that the diplosomes enter separately into the equatorial plate of the first maturation monaster.

6. Nezara hilaris Say.

Spermatogonic Divisions. — In the equatorial plate (Plate IX, Fig. 38) there are 14 chromosomes; 12 are autosomes that compose 6 pairs of gradated volumes (A, a–F, f), while the two smallest are apparently not quite equal in volume (Di, di) and are the diplosomes.

Growth Period. — The diplosomes conjugate and remain in close contact during the growth period (Fig. 39, Di, di). From the late synapsis stage on each appears plainly constricted, which is probably to be interpreted as a longitudinal splitting.

There were no later stages upon my slides.

Literature. — In the former paper (1901b) I was mistaken in supposing there to be 16 chromosomes in the spermatogonia. Wilson (1905a) presents observations upon the later stages, and shows that the diplosomes divide separately and equationally in the first maturation division, but conjugate and separate reductionally in the second; but he is mistaken in saying that the diplosomes are of equal volume.

7. Brochymena sp.

Spermatogonic Division. — Pole views of the equatorial plate (Plate IX, Figs. 40, 41) show 14 chromosomes, of which 12 (A, a–F, f) form 6 pairs of graduated volumes in which the two members of each pair are approximately equal in form and volume; while the remaining pair consists of one element (Di) that is the smallest of all and of another (di) that is constricted and is larger than either of the components of the autosome pair, F, f.

Growth Period. — The twelve autosomes unite to form 6 bivalent ones. The diplosomes also conjugate, and each becomes constricted as in *Euschistus variolarius*.

First Maturation Division. — Late in the prophase the diplosomes separate and enter into the chromosomal plate apart from each other (Di, di, Figs. 42, 43). These divide equationally, but the 6 bivalent autosomes reductionally.

Second Maturation Division. — Here there are 6 univalent autosomes that divide equationally (Figs. 44, 45, *A–F*). But the diplosomes conjugate in the center of the equator and this bivalent element (*Di, di*), with components of very unequal volume, divides reductionally. Accordingly each spermatid receives 6 autosomes and one of the two diplosomes.

This is another species where the particular chromosome pairs may be recognized with great precision in each cell generation, as one finds by comparing the correspondingly lettered elements in the figures.

Literature. — I previously (1901*b*) concluded there were 16 instead of 14 chromosomes in the spermatogonia, for I was misled into counting two constricted elements as two each. Further I did not notice that the diplosomes enter separately into the plate of the first maturation mitosis, and did not describe the following mitosis. Wilson (1905*a*) described and figured this process correctly.

8. PERILLUS CONFLUENS H.-S.

Spermatogonic Divisions. — There are 14 chromosomes (Plate IX, Fig. 46) of which 12 form 6 gradated pairs of autosomes (*A, a–F, f*); while the two smallest elements (*Di, di*) are not of quite equal volume and are diplosomes as the later history shows.

Growth Period. — Six bivalent autosomes are formed. The diplosomes also conjugate but later in the synapsis stage than in the other Pentatomids. Subsequently each becomes constricted, and they lie close together and at the same time against the plasmosome (Fig. 47).

First Maturation Division. — In the late prophases the diplosomes separate and lie in the chromosomal plate near each other (Fig. 48, *Di, di*); each divides through the plane of its previous constriction. Fig. 49 shows a daughter chromosomal plate of the early anaphase of this mitosis; 6 show a line of division and they are univalent autosomes with the reopening longitudinal split, while the two that show no such constriction are the autosomes.

Second Maturation Division. — On pole view of the spindle (Fig. 50) are seen 7 elements of which the central one is really bivalent, formed by the conjugation of the two univalent diplosomes (*Di, di*). Fig. 51 represents a lateral view of the same stage but showing only 6 of the 7 elements; the one with the two components of unequal volume is the bivalent diplosome. This diplosome divides reductionally, the autosomes equationally; consequently each spermatid (Fig. 52) receives 7 elements, namely, 6 autosomes and one of the two diplosomes.

Literature. — My previous description was erroneous in stating there to be 16 chromosomes in the spermatogonia, and in failing to note that the diplosomes lie separate in the first maturation monaster. I did not describe the second maturation mitosis.

9. CŒNUS DELIUS Say.

Spermatogonic Divisions. — In the equator of the spindle there are 14 chromosomes (Plate IX, Figs. 53, 54). Ten of these compose 5 pairs of gradated sizes, each pair with components of equal volume (A, a–E, e). Of the remaining 4 I take 2 (F, f) to be another pair of autosomes, though they are not quite equal; while 2 others still more unequal in size (Di, di) are probably the diplosomes judging from the later history of the chromosomes in the spermatocytes. That all of these elements become halved in the anaphase is shown by the recurrence of the number 14 in a daughter chromosomal plate (Fig. 55).

Growth Period. — The two very unequal diplosomes may be either united during the growth period, which appears more frequent, or they may be separated.

First Maturation Division. — Eight chromosomes enter into the spindle, and were all shown on lateral view in Fig. 61 of my earlier paper (1901b). They are 6 bivalent autosomes that divide reductionally, and 2 separated diplosomes that divide equationally. A pole view of a daughter chromosomal plate of the early anaphase is shown in Fig. 56; the 6 bipartite elements are univalent autosomes with the reopening longitudinal split, and the two unipartite bodies in the center are the diplosomes (Di, di).

Second Maturation Mitosis. — The two diplosomes conjugate in the center of the equatorial plate (Figs. 57, 58), and in the anaphase separate from each other without dividing, while the 6 autosomes divide equationally.

Literature. — My previous account (1901b) was incorrect in stating 16 to be the number of spermatogonial chromosomes, and in considering the diplosomes to divide reductionally in the first maturation mitosis; then I did not follow the spermatogenesis beyond this point. Wilson has given a full account of the whole process, and my present observations corroborate his in every particular, except that I find the two diplosomes to be by no means always regularly separated from each other in the growth period as Wilson describes.

10. TRICHOPEPLA SEMIVITTATA Say.

Spermatogonic Divisions. — Fig. 59, Plate IX, is a careful redrawing of the chromosomal plate illustrated in Fig. 65 of my earlier paper (1901b). It shows distinctly 15 elements, while the small protuberance Z attached to the chromosome a may be a sixteenth. From the phenomena of the growth period there are to be concluded at least 16 chromosomes for the spermatogonium, in agreement with my former description. Twelve, which compose a series of gradated pairs (A, a–F, f), are probably autosomes, while two remaining elements of very unequal volume (*Di, di*) are probably correspondent to the two larger diplosomes of the later stages. The minute body lettered *Z* is probably another diplosome and so also the one lettered *Y*. All the chromosomes are characterized by rather uneven and irregular outlines.

Growth Period. — Twelve autosomes unite to form 6 bivalent ones as shown by the phenomena of the subsequent prophases. The two larger diplosomes (*Di, di*, Figs. 60–63) usually lie close together in the earlier growth period, but separate from each other either soon after or else not until the late prophases. When in contact their long axes may be parallel, but more usually they are crossed. At an early stage each becomes distinctly split along its length, but this usually closes soon after it becomes well marked, which is associated with the phenomenon that each diplosome swells in size and becomes more spherical ; just before the following mitosis this split reappears on each as a transverse constriction. Besides these two larger diplosomes more minute ones are to be seen during the growth period, and despite their small size may be easily distinguished by their deep stain from the pale autosomes. It is very difficult to decide exactly what their number is, though in most cases 3 or 4 such bodies can be found. Generally two minutest ones of equal volume (*K*, Figs. 61, 63) lie upon the surface of the largest plasmosome (*Pl*), while 1 or 2 slightly larger ones (*x*, Figs. 62, 63) are situated elsewhere in the nucleus and sometimes in contact with smaller plasmosomes. The 2 smallest, those upon the largest plasmosome designated by the letter *K*, are always close together and of equal size, therefore they are probably (longitudinal?) division products of a single one ; while the two others are usually widely separated and of unequal size. These four smallest diplosomes of the growth period may be represented by three minute elements in the spermatogonium : we found in that stage (Fig. 59) one minute element (*Y*) and another probably separate element (*Z*), and there might be still another in this chromosomal plate but hidden from view. Accordingly, judging from the phenomena of the growth period, there must be at least 4 diplosomes represented in the spermatogonium, that is, a total of 16 chromosomes, if not indeed 5 diplosomes.

First Maturation Mitosis. — There are always at least 8 distinct elements in the spindle, which are : 6 bivalent autosomes of very different volumes (*A, a–F, f*, Fig. 65) which undergo a reduction division ; and two univalent diplosomes (*Di, di*) which divide presumably equationally, and represent the diplosomes so lettered in the preceding stages. The minute diplosomes are rarely found in the equatorial plate, but in two cases, one of them shown in Fig. 64, a pair of small bodies (*x*) placed close together were found ; they do not appear to divide with the other chromosomes and seem afterwards to move out into the cytoplasm ; they may represent the small elements marked K and x of Figs. 61–63, and the elements Z and Y of the spermatogonium (Fig. 59).

Second Maturation Division. — On pole view of the spindle (Plate X, Fig. 67) are seen 7 chromosomes, the central one of which is bivalent and represents the two larger diplosomes placed end to end as lateral views evince (Fig. 66, *Di, di*); this bivalent chromosome divides reductionally, the 6 autosomes probably equationally. In the spermatids (Fig. 68) there are always 7 chromosomes, half of the spermatids containing the larger and half the smaller component of the larger diplosome pair.

Literature. — My previous account was entirely correct, except that I failed to note that the larger diplosomes divide equationally in the first maturation mitosis. Wilson (1905a) described the second maturation mitosis correctly, but could not follow the history of the smallest diplosomes any more satisfactorily than I have been able to do in either of my accounts.

11. EURYGASTER ALTERNATUS Say.

Growth Period. — There are two diplosomes of very different volumes (*Di, di*, Plate X, Fig. 69); this figure shows also three whole bivalent autosomes. In the earlier period these are usually, not always, placed end to end. Each is at first elongate, in the postsynapsis undergoes a split through its length, and for a considerable time retains this fissure in this position; later each half of each diplosome rounds up so that the whole appears to be transversely constricted, but this constriction is the same as the earlier split. There is no complete rest stage.

First Maturation Division. — There are always 7 chromosomes (Fig. 70); the two smallest (*Di, di*) are the diplosomes that come to lie separately in the equator and divide equationally; their precise location in the chromosomal plate is variable. The others are 5 bivalent autosomes that divide reductionally as may be ascertained with great certainty from the examination of the earlier stages; and when seen from the flat surface each shows the longitudinal split parallel to the long axis. In the succeeding anaphase this split opens up as in the other Hemiptera.

Second Maturation Mitosis. — Pole views (Fig. 72) show apparently only 6 chromosomes, but the central one is really bivalent, composed of the two diplosomes (*Di, di*) placed end to end; a lateral view shows this bivalent element more distinctly (Fig. 73). The diplosomes divide reductionally, the autosomes equationally, so that each spermatid receives 6 elements.

Though there were no spermatogonic mitoses upon my preparations, there can be little doubt that the chromosomes there would consist of 10 autosomes and 2 diplosomes.

Literature. — My previous very brief account was correct so far as it went.

12. PERIBALUS LIMBOLARIS Stal.

Spermatogonic Divisions. — There are 14 chromosomes (Plate X, Fig. 74) ; 12 of them make up 6 well marked pairs of autosomes (*A, a–F, f*), and all of these are elongate ; the two remaining are very unequal in volume (*Di, di*), are rounded, are the smallest of all, and are the diplosomes. The gradation in size of the autosome pairs is very marked.

Growth Period. — During the greater part of the growth period there appears to be only one diplosome in the spermatocytes, and it usually is of rounded form and contains one or several vacuoles ; whether this single one represents both diplosomes of the spermatogonia, or only the larger one of them, I could not positively determine. Towards the close of this period, however, two separated ones of very dissimilar volume are occasionally found (Fig. 75, *Di, di*). During the synapsis, unlike the conditions in the other Pentatomids, these are not safraninophilous but stain violet like the plasmosomes of which there are usually two or three in each nucleus, and for this reason it is then difficult to determine the diplosomes.

First Maturation Mitosis. — In the equator of the spindle are present always 8 chromosomes (Figs. 76, 77) ; the two smallest are the diplosomes which have entered the spindle separately and divide there equationally ; they are dyads. The 6 larger elements are bivalent autosomes, each of which appears as a tetrad with distinct components when seen from its flattened surface (Fig. 77) ; the longitudinal split of these is parallel to their long axes, the same position as it held in all the earlier stages, and accordingly in this first maturation mitosis the autosomes divide reductionally. A pole view of one of the daughter chromosome plates, from the early anaphase, is illustrated in Fig. 79 ; the diplosomes (*Di, di*) can be readily distinguished from the autosomes by being unipartite and smaller.

Second Maturation Division. — Pole views show apparently only 7 elements (Fig. 78) ; but the central one is seen to be composed of two placed the one immediately above the other (*Di, di*), which are the now conjugated diplosomes. This bivalent diplosome is more easily recognized upon side view (Fig. 80), and divides reductionally, *i. e.*, the larger diplosome (*di*) passes into one spermatid and the smaller diplosome (*Di*) into the other, while the 6 autosomes divide through the plane of their longitudinal splits.

Literature. — I had erroneously (1901*b*) stated the number of spermatogonial chromosomes to be 16, and was consequently led into concluding that there is a bivalent diplosome dividing reductionally in the first spermatocyte division.

NABIDÆ.

13. NABIS ANNULATUS Reut.

On my preparations there were no stages of the spermatogonia or earlier portion of the growth period.

First Maturation Mitosis. — Very early prophases show 6 autosomes in the form of long loops which are evidently to be considered tetrads with a very wide longitudinal split. Besides these there is apposed to a plasmosome (*Pl*, Plate X, Fig. 81) a still larger body (*Di*), safraninophilous, of uneven contours, which the later history shows to be a number of allosomes in close juxtaposition. Later the 6 autosomes shorten and condense, and then each appears to consist of two parallel univalent elements each longitudinally split, as illustrated by those marked *m* in Figs. 81–83; each of these gradually condenses into a tetrad composed of four parallel rods, whereas in most other Hemiptera the univalent elements come to lie end to end; further, the longitudinal split remains open instead of closing temporarily. In these later prophases the safraninophilous body (*Di*, Fig. 81) separates into 4 allosomes, while the plasmosome to which it is attached gradually dissolves (Figs. 82, 83). Two of these compact allosomes are quadripartite (*Di. 2*), and each of these is therefore probably, and the later history confirms this decision, a bivalent, longitudinally split chromosome; these are the ones lettered *Di. 2, di. 2* and *Di. 3, di. 3* in Figs. 82, 83 and 85. Each is, that is to say, a bivalent diplosome with its components in close contact and with these components of approximately equal volume. But the remaining pair of allosomes consist of the largest and the smallest respectively, and are very unlike in volume, while each is a dyad and not a tetrad (*Di. 1, di. 1*, Figs. 82–85). These relations cannot be determined as long as these bodies are in close contact, but very clearly as soon as they become separate. These three pairs of diplosomes are readily distinguished from the autosomes by their dense and rounded form and their strong affinity for the safranine stain. There are accordingly three pairs of diplosomes in the spermatocyte, two of them tetrads, and one pair with widely separated components of unequal volume.

Pole views of the first maturation monaster show always 10 chromosomes (Fig. 86). Eight of these are clearly quadripartite, as can be readily determined when the pole view is slightly oblique as that of the figure given, and these must correspond to the 8 tetrads of the prophases, namely, to the 6 bivalent autosomes, and to the 2 bivalent diplosomes marked *Di. 2, di. 2* and *Di. 3, di. 3;* which two, however, are these particular diplosomes, cannot be determined with certainty in the stage of the equatorial plate. The two remaining elements are not tetrads but dyads, they are of unequal volumes (*Di. 1, di. 1*, Figs. 86–88), and clearly represent the third pair of diplosomes of the preceding prophases; they are respectively the largest and the smallest elements of the chromosomal plate. Each tetrad is composed of 4 parallel rods, shown in their length in Fig. 86, and from end in Figs. 87, 88; their long axes always lie in the plane of the equator. But in the case of the two dyads, the larger (*di. 1*) may have its long axis in this plane (Fig. 88), but more frequently is inclined to it (Fig. 87); while the smaller dyad (*Di. 1*) is composed of two spherules, one on either side of the equatorial plane. All these chromosomes are large, and their parts can be made out with unusual facility. Each of these 10 elements divides so that each second spermatocyte receives 10, *i. e.*, a portion of each of them. Whether this is a reductional or an equational division of the 8 tetrads it would be exceedingly difficult to determine, since each, as in the case of *Ascaris*, is in the form of four parallel rods; but I conceive that these 8 bivalent elements differ from those of other Hemiptera only in having their univalent components placed side to side instead of end to end, and that therefore their division may well be, as is certainly the case in the other Hemiptera, reductional. A pole view of one daughter chromosomal plate in the early anaphase is shown in Fig. 89; here are 8 bipartite elements, the daughters of the former 8 tetrads, and 2 unipartite ones (*Di. 1, di. 1*), the division products of the 2 earlier dyads.

Second Maturation Mitosis. — The 8 bipartite elements, which are 6 autosomes and 2 of the diplosomes, take positions with their long axes in the plane of the equator (Figs. 90, 91), and all of them divide so that the components of each become separated into opposite spermatids; this is probably an equational division. But the unipartite diplosomes *Di. 1* and *di. 1* never lie in the equator, but one is always near one spindle pole and the other near the opposite pole; this was invariably the case with every one of these stages found. Accordingly, the smaller diplosome, *Di. 1*, passes wholly into one spermatid, the larger diplosome, *Di. 1*, into the other spermatid. Fig. 92 shows the chromosomes of a spermatid that has received the smaller one, and Fig. 93 a spermatid that has gotten the larger, these diplosomes being recognizable among the other chromosomes by their form as well as by their deeper stain.

In the spermatocytes there are accordingly 6 autosomes that divide in both maturation mitoses; 2 probably bivalent diplosomes each of which divides as do the autosomes; but one pair of diplosomes, that one characterized by very unequal components, each component dividing separately (so probably equationally) in the first mitosis, but their daughter products, without conjugating, passing without division into opposite spermatids in the second mitosis.

The 6 quadripartite autosomes are probably, by analogy with the phenomena of the other Hemiptera, bivalent in the spermatocytes, and so are probably the 2 quadri-partite diplosomes; the large and small diplosomes are undoubtedly univalent. Therefore we can postulate for the spermatogonium with a high degree of certainty: 12 autosomes, and 6 diplosomes, the components of only one of these diplosome pairs being very unequal in volume.

Literature. — My preceding account (1901*a*), which did not extend beyond the first maturation mitosis, was entirely correct except for the conclusion that the sper-matocyte had four bivalent diplosomes. My preparations of *Coriscus ferus*, another member of the same family, had faded to such a degree that I could not test the cor-rectness of my account of it (1901*b*).

COREIDÆ.

14. Harmostes reflexulus Say.

Spermatogonic Divisions. — There are 13 chromosomes. One unpaired element (Plate X, Figs. 94, 95, *Mo*) is the monosome, and it is not the largest. The 2 smallest are the diplosomes (*Di, di*) and are not quite equal in volume. The remain-ing 10 are autosomes and are seen to compose 5 readily recognizable pairs (*A, a–E, e*); what is to be noted in them is that the two components of each pair seem to be of slightly different form and volume, as is seen most clearly in the case of the pair *A, a*; and perhaps in each pair the larger element may be the maternal one and the smaller the paternal. The components of the 2 or 3 largest pairs are regularly transversely constricted.

Growth Period. — The 10 autosomes conjugate to form 5 bivalent ones. The monosome (*Mo*, Figs. 96–99) remains safraninophilous during this whole period. In the synapsis (Fig. 96) it becomes elongated and concomitantly more or less bent, thereby showing a great variety of forms; frequently it is attenuated at the ends and thicker at the middle. In the early postsynapsis (Fig. 97) it becomes longitudinally split so that the halves sometimes widely diverge from each other and at the same time it becomes less dense and more or less granular, though to much less extent than the autosomes (Fig. 98). In the rest stage, which is complete (Fig. 99), this split becomes more or less closed; and then the monosome (*Mo*) has usually a rod shape, shorter than in the synapsis stage, with its arms parallel; throughout the growth period it lies against the nuclear membrane. I could not distinguish the diplosomes in the earlier part of the growth period before the plasmosome arises. In the rest stage the latter (*Pl*, Fig. 99) is a large body near the center of the nucleus. Quite generally there are attached to its surface about 3 or 4 small safraninophilous bodies; the 2 larger that may or may not be in contact I take to be the diplosomes (*Di, di*);

the smaller ones (x) are bodies represented in neither the spermatogonic nor the spermatocytic mitoses. In the case figured (Fig. 99) the bivalent diplosome has each component longitudinally split.

First Maturation Division. — In the early prophases (Figs. 100, 101) a bivalent diplosome (Di, di) is frequently to be seen lying near the monosome (Mo), which might indicate that previously it had been in contact with it, from which it would appear possible that when the diplosomes are not discernible in the preceding rest period it is because they may be closely applied against the monosome. The diplosomes seem not to increase in size during the growth period. In these prophases the longitudinal split of the monosome again appears.

In the chromosomal plate (Figs. 102, 103) there are always present 1 bivalent diplosome (Di, di) that divides reductionally, and 1 monosome (Mo) that divides through the plane of its longitudinal split. There may be either 5 bivalent autosomes (Fig. 102, A, a–E, e) all of which divide reductionally; or 4 bivalent autosomes (A, a–C, c, E, e, Fig. 103) and 2 univalent ones (D, d); in the latter case the 2 univalent ones are regularly of the same form and volume, and therefore are evidently ones that had either failed to conjugate or, more probably, ones that had precociously separated from each other after conjugation, and which in this mitosis pass without division into opposite daughter cells, *i. e.*, divide reductionally as do the other autosomes. The longitudinal split is well marked upon one or two of the larger autosomes.

Second Maturation Division. — Here there are always 7 elements (Fig. 104, where one of the autosomes has not yet taken its place in the equator of the spindle). The smallest, the diplosome (Di), regularly divides, and so do the 5 autosomes, all of these equationally. But the monosome (Mo) shows no sign of any division and passes bodily over into one of the spermatids. The latter show correspondingly either 6 chromosomes (Fig. 105) or 7 (Fig. 106), the monosome being absent in the former case; the minute element in each spermatid is a diplosome.

Literature. — My preceding accounts (1901a, b) were correct in the main, stated the spermatogonial number of chromosomes accurately, the variation in number in the first maturation spindle, and the behavior of the monosome in the maturation divisions. But what escaped me then was that the large allosome of the growth period is the monosome and not the bivalent diplosome.

15. CORIZUS ALTERNATUS Say.

Spermatogonic Divisions. — There are 13 chromosomes (Plate X, Fig. 107). The smallest elements, of slightly different volume, are the diplosomes (Di, di). Then 5 pairs of autosomes (A, a–E, e); of these the largest pair (A, a) is composed of 2 relatively enormous elements, one of which is approximately straight and apparently a little more voluminous, while the other is horseshoe-shaped. Finally there is a single chromosome without a corresponding mate, therefore a monosome (*Mo*).

Growth Period. — In the synapsis stage the 10 autosomes become longitudinally split and conjugate to form 5 bivalent ones. But 3 of the chromosomes differ in preserving their safraninophilous stain and dense structure; from the later history of these there can be no question that the largest (*Mo*, Figs. 108–111) is the monosome, the 2 smaller the diplosomes (*Di, di*). The monosome increases somewhat in volume and in the postsynapsis (Figs. 109, 110) is rod-shaped, sometimes bent, and undergoes a longitudinal splitting; in the rest stage, that is complete (Fig. 111), it becomes more rounded and then shows either no trace of this split, or else only a mere sign of it in the form of an indentation at either end; it may or may not lie against the nuclear membrane. The diplosomes are unequal in volume as in the spermatogonium, and undergo but slight increase in mass during the growth period. In the postsynapsis each (*Di, di*, Fig. 109) becomes bipartite, which is evidently a longitudinal splitting, and they remain so during the remainder of the growth period. The spermatocytes contain each several large plasmosomes (*Pl*, Figs. 110, 111), and the diplosomes, and less frequently the monosome, may be in contact with these.

First Maturation Division. — In the prophases there are 5 bivalent autosomes (*A, a–E, e*, Figs. 114–116), each longitudinally split. One of them, by far the largest (*A, a*), is in the earlier stages the single one that is regularly ring-shaped (Fig. 112), with a distinct longitudinal split in each arm of the ring; this ring gradually opens until it first becomes an angle (Fig. 113), then straight (Figs. 114–116), the longitudinal split still continuing in the axis of each arm (univalent constituent). By the gradual condensation of the autosomes (Fig. 116) their longitudinal splits become more or less closed, but even in the metaphase it is sometimes clearly indicated (Plate XI, Fig. 118), and is then always parallel to the long axis of the chromosome. No animal shows more decisively than this one that the first maturation mitosis separates whole univalent chromosomes. The monosome can be recognized as a large dyad (*Mo*, Figs. 114–116). The diplosomes (*Di, di*, Figs. 114–116) do not conjugate until the later prophases, apparently usually not until the nuclear membrane has disappeared; in them the longitudinal split becomes temporarily closed as in the case of the autosomes, but the monosome continues to show it distinctly.

There are in the spindle almost invariably 7 elements (Plate XI, Figs. 117, 118); in a few cases 8 are to be seen on pole aspect, which is then due, as in *Harmostes*, to a precocious division of two of the bivalent elements, but here usually of the bivalent diplosome. There is a central bivalent diplosome (*Di, di*) and around it a circle composed of 5 bivalent autosomes and the univalent monosome (*Mo*, Fig. 117); the latter can be recognized on pole view by its lesser depth, and on lateral view (Fig. 118) by its quadratic form. The constrictions of the autosomes seen on pole view mark their longitudinal splits, as is very clearly proven by the earlier history of these chromosomes. The bivalent diplosome and autosomes divide reductionally, the monosome equationally. Fig. 119 reproduces a daughter plate of chromosomes from the early anaphase; the monosome (*Mo*) can be recognized as being the only element that shows no longitudinal split.

Second Maturation Division. — Here again there are always 7 elements (Plate XI, Figs. 120, 121), the smallest being a diplosome (*Di*), and the one that is rounded without having any constriction the monosome (*Mo*). The diplosome and the 5 autosomes always divide, but the monosome passes wholly over into one of the spermatids; this is shown clearly by the anaphase shown in Fig. 122, where at one spindle pole are 7 elements and at the other only 6.

Literature. — My preceding description (1901*a*) was incorrect in giving 14 as the normal number of chromosomes; this was because I had counted into the chromosomal plate elements of an adjacent cell. Further, I had entirely overlooked the presence of a monosome, and had not described the second maturation mitosis.

16. CORIZUS LATERALIS Say.

No spermatogonic divisions were found.

Growth Period. — My preparations had faded considerably so that I could not make out the diplosomes with any certainty. But the largest allosome present is the monosome and it becomes longitudinally split.

First Maturation Division. — There are 7 elements (Plate XI, Fig. 123): 5 bivalent autosomes and 1 bivalent diplosome (*Di, di*), with components of dissimilar volume) that divide reductionally; and 1 roundish element, the monosome (*Mo*), that also divides but equationally.

Second Maturation Division. — Again 7 elements: 5 autosomes and 1 diplosome (*di*) that divide again, and a rounded monosome (*Mo*) that passes into one spermatid without division, as shown in all lateral views of the anaphase (Fig. 125).

The whole spermatogenesis seems very similar to that of the preceding species, and we may conclude with considerable certainty that there will be found in the spermatogonia: 10 autosomes, 2 diplosomes and 1 monosome.

Literature. — My earlier account (1901*b*) was in the main correct, and though I did not decide for the presence of a monosome I noted that one of the chromosomes of the first maturation mitosis differed in form from the others, " for it is not more than half the volume of the other five, and sometimes it does not appear dumbell-shaped."

17. CHARIESTERUS ANTENNATOR Fabr.

There were no spermatogonic divisions suitable for study.

Growth Period. — In the synapsis and later stages (a complete rest stage was not observed) there are in each nucleus two compact, safraninophilous bodies, close to the nuclear membrane; a plasmosome was not found. The smaller of these bodies (*Di, di,* Plate XI, Fig. 126) is regularly constricted, and by analogy with the relations in other members of the family is probably a bivalent diplosome, and its later history is in accord with this assumption. The larger safraninophilous body is longitudinally split (*Mo*), and corresponds to the monosome of the later stages.

First Maturation Division. — Pole views of the chromosomal plate show in most cases (14 out of 18) 13 elements (Fig. 127). The central is always the smallest, and very likely is a bivalent diplosome (*Di, di*); its two components are of approximately the same size. Around it is a circle of 11 autosomes, and just outside of the latter an element (*Mo*), the monosome, lying with its long axis in the equator while the autosomes are perpendicular to it. In 4 out of the 18 clear pole views examined there appeared to be 14 elements (Fig. 128); these are to be interpreted, as in *Harmostes,* that one of the bivalent autosomes has its univalent components precociously separated; and in all such cases illustrated by Fig. 128 there lie near each other two elements of equal volume (*M*), each of which is of less depth than any other of the autosomes. The autosomes and the diplosome divide reductionally, the monosome through the plane of its longitudinal split (Fig. 129).

Second Maturation Division. — Here there are always 13 elements (Fig. 130). The smallest is a diplosome (*di*), 11 others are autosomes, and all these divide equationally. But the monosome passes without division into one of the spermatids. This is shown distinctly in two daughter chromosomal plates of the early anaphases of the same cell, the drawings made accordingly at different focusses (Figs. 131, 132); in each there is a diplosome recognizable by its very small size, but only one shows the monosome (*Mo,* Fig. 131). And in later anaphases on lateral views (Fig. 133) are to be seen regularly an element, the monosome, in one spermatid that is not found in the other. Half the spermatids receive, accordingly, 13 elements, and half 12.

Judging from the relations during these maturation mitoses the number of chromosomes in the spermatogonia would be: 1 monosome, 2 diplosomes, 22 autosomes, a total of 25.

Literature. — My preceding observations (1901*b*) were correct, and though I did not distinguish a monosome in the growth period of the spermatocytes, I called attention to the fact that one of the chromosomes of the first maturation mitosis is different in form from the others, and left the question open whether it might be univalent there (so be a monosome). The subsequent mitosis was not described.

18. PROTENOR BELFRAGEI Hagl.

The previous account given by me (1901b) was detailed and entirely correct, and Wilson has recently corroborated it. I have simply to add to it that all the autosomes of the spermatogonium can be grouped into pairs (A, a–E, e, Plate XI, Fig. 134), that the diplosomes there are slightly unequal in volume (Di, di), and that the monosome (Mo) is by far the largest element. Another figure (135) is given of these elements in the growth period. The monosome becomes always longitudinally split in the synapsis period (Mo, Fig. 135), and its division in the first maturation mitosis is along the plane of this split and not, as l had previously interpreted it, transverse to its long axis.

19. ALYDUS PILOSULUS H. S.

Spermatogonic Division. — Four clear pole views showed in each case 13 elements, namely (Plate XI, Fig. 136): 5 pairs of autosomes A, a–E, e of remarkably different volumes and forms; 2 unequal diplosomes (Di, di), the smallest of all; and 1 monosome (Mo).

Growth Period. — In the growth period there is a single safraninophilous body of considerable size, that from its singularity and later behavior is undoubtedly the monosome (Mo, Figs. 137, 138), and from the early synapsis on increases to at least twice its original volume, as shown by comparison of the figures. In the postsynapsis it becomes longitudinally split, lies regularly against the nuclear membrane and frequently also against a plasmosome. The diplosomes are apparently not distinguishable during the growth period, and therefore it is probable that they undergo much the same changes as the autosomes except for their later conjugation.

First Maturation Division. — In the prophases the diplosomes (Di, di, Fig. 139) become compact ahead of the autosomes, and reappear as two rounded bodies that do not conjugate until the nuclear membrane disappears. The monosome (Mo) is to be distinguished from them by its larger size. The autosomes are longitudinally split and bivalent. In the equatorial plate (Fig. 140) there are always 7 elements: 5 bivalent autosomes that divide reductionally, and a bivalent diplosome (Di, di) that divides in the same manner as may be readily determined on the basis of its two components being dissimilar in volume. The monosome (Mo) divides lengthwise. The bivalent diplosome is always central, the monosome most excentric. In a number of cases two of the larger autosomes were found closely applied side to side and in the preceding late prophases this is also sometimes the case.

Second Maturation Division. — Again 7 elements are found (Fig. 141), the smallest of which is the diplosome, the nonconstricted one the monosome (*Mo*). All of these divide except the monosome which passes wholly over into one of the spermatids, as shown clearly in the anaphase illustrated in Fig. 142 where one daughter plate shows 7 and the other only 6 elements. The monosome frequently lags behind the others in reaching the spindle pole (Fig. 143).

Literature. — My preceding account (1901*b*) was very brief, I overlooked the monosome entirely and erroneously gave 14 chromosomes as the normal number. Wilson (1905*c*, 1906) has correctly emended my observations and has given a good series of figures, but he failed to note that the diplosomes are unequal in size.

20. ALYDUS EURINUS Say.

My earlier accounts (1901*b*, 1905 p. 194) were correct, except that I failed to note that the allosome of the growth period (*Mo*, Plate XI, Fig. 145) is the odd chromosome, *i. e.*, the monosome, and not a bivalent diplosome; there is no trace during the growth period of the very minute diplosomes. The monosome is rather ovoid in the synapsis period, but it later becomes more elongate and longitudinally split (this split shows usually simply as an indentation at either end, but sometimes as a fine clear line along the whole length). Its division in the first maturation mitosis (Fig. 147) is in the line of this split, therefore equational. A daughter chromosomal plate of this division is reproduced in Fig. 148; the monosome is the only element that appears unconstricted, while all the others, including the small central diplosome (*Di*), show a constriction that is the longitudinal split reopening for the next mitosis. In the second mitosis there are again 7 elements, all of which divide except the monosome (*Mo*) that passes without division into one of the spermatids. In the spermatogonium (Fig. 144) the 13 chromosomes make up 5 pairs of autosomes (*A, a–E, e*) one pair of diplosomes (*Di, di*), and the monosome (*Mo*). The whole spermatogenesis is quite similar to that of the preceding form.

21. ANASA TRISTIS De Geer.

Spermatogonic Divisions. — In seven very clear pole views 21 chromosomes could be counted. These are (Plate XI, Fig. 151): 2 small rounded bodies, not quite equal in size, the diplosomes (*Di, di*); a longest unpaired one that is sometimes constricted, the monosome (*Mo*); and a series of 9 pairs of autosomes (*A, a–I, i*).

Growth Period. — The large allosome of the growth period is the monosome (*Mo*, Figs. 152–155), which remains compact and safraninophilous. It is irregularly elongate during the synapsis (Fig. 152) and in the later postsynapsis (Fig. 155) shows a split along its length which, as is the case also with the autosomes, is widest at its middle; this split becomes temporarily closed a little later. The diplosomes (*Di, di,* Figs. 153, 154) remain very small during the growth period but retain their red stain and dense structure; usually but not always they are close together, and like the monosome lie against the nuclear membrane. There is always one large plasmosome (Figs. 154, 155, *Pl*) and frequently one or two smaller ones.

First Maturation Mitosis. — In the spindle there are 11 elements so placed that within a circle of 9 autosomes is the bivalent diplosome (*Di, di,* Fig. 156), and outside of this circle the univalent monosome (*Mo*) which lies with its long axis in the equatorial plane; the annular constrictions of the autosomes found upon pole views mark their longitudinal splits. All of these are shown on lateral view in Fig. 157, and 6 of them in Fig. 158. The 9 autosomes divide reductionally, and so does the bivalent diplosome because its parts that separate from each other are unequal in volume and in the preceding stages we found this dissimilarity characteristic of the two. The monosome, however, lies with its long axis in the plane of the equator (Figs. 157, 158, *Mo*), and divides through its length.

Second Maturation Division. — Here again there are 11 elements (Fig. 159), but grouped differently from those of the preceding division in that there are usually 2 within a circle of 9. They are 1 univalent diplosome (*Di*), 9 univalent autosomes, and the half of the monosome. The autosomes and the diplosome divide again and equationally (Fig. 160), but the monosome (*Mo*, Figs. 160, 161) passes undivided into one of the spermatids and usually lags behind the others in reaching the spindle pole.

Literature. — Paulmier's monographic account of the spermatogenesis of this species (1899) was in the main a very correct one, save that he stated the normal number of chromosomes to be 22, and consequently identified the allosome of the growth period and the chromosome that does not divide in the second maturation mitosis with the minute diplosomes. I (1901*b*) followed Paulmier in these mistakes, and because the monosome of the spermatogonium is constricted counted it as two. Wilson (1905*c*, 1906), in whose laboratory Paulmier's work was done, was the first to correct these errors, and to trace the history of the monosome distinct from that of the diplosomes. But Wilson failed to note that the diplosomes are not quite of the same size, and that they may be distinctly recognized during the greater part of the growth period.

22. ANASA sp. (from California).

Spermatogonic Divisions. — In every case there are 21 elements in the spindle (Plate XI, Fig. 164). These are: 2 diplosomes of unequal volume (*Di, di*); 1 monosome that appears to be regularly constricted (Mo); and 9 pairs of autosomes ($A, a-1, i$).

Ovogonic Divisions. — On the only two clear pole views upon my preparations there were exactly 22 elements.   A careful comparison shows that the odd one of the spermatogonia, the monosome (Mo, Fig. 164), is represented in the ovogonia (Figs. 162, 163) by a pair of elements ($Mo, \overline{\ } mo$); each component of this ovogonic pair is of about the same volume as the single monosome of the spermatogonia.   In the ovogonia there are also a pair of diplosomes of dissimilar volumes.

Growth Period. — The monosome and the diplosomes show the same behavior as in the preceding species, and the longitudinal split of the monosome is very distinct.

First Maturation Division. — Pole views show 11 elements, in the center the bivalent diplosome (Di, di, Fig. 165) and a bivalent autosome, then a circle of 8 bivalent autosomes, and outside of the latter the monosome (Mo).   All of these divide reductionally except the monosome (Mo, Fig. 166) that divides equationally.

My preparations contained no second maturation mitoses, but probably the monosome will be found to behave in them as it does in *Anasa tristis.*

Literature. — My earlier account (1901*b*) was erroneous in stating the spermatogonic number of chromosomes to be 22; because the monosome there is regularly constricted I was misled into counting it as two.   And that led to the further mistake of concluding the allosome of the growth period to be the bivalent diplosome.

23. Anasa armigera Say.

Spermatogonic Divisions. — On the only two clear pole views of chromosomal plates 21 elements could be counted (Plate XI, Fig. 167); here the monosome is the only one that is somewhat constricted (Mo) and is not the largest; then there are 2 very small diplosomes (Di, di) of nearly equal size, and 9 pairs of autosomes ($A, a-I, i$).

Growth Period. — The staining of my single preparation was not favorable for determining the behavior of the diplosomes, but the large allosome must be the monosome on account of its similarity to that of the other species of this genus.

First Maturation Division. — There are 11 elements, all shown in Fig. 168.   The smallest is the bivalent diplosome (Di, di), while the monosome can be recognized by its unipartite appearance (Mo).   I have seen stages no later than this metaphase, but it is sufficient to show that the autosomes and the diplosomes divide reductionally.

Literature. — My previous very brief account (1901*b*) made the same mistakes as I had made for the other species of the genus.   In the figure then given of the spermatogonic chromosomes (Fig. 77, 1901*b*) I had counted the constricted one just to the left of the two diplosomes as two whereas it is really but a single monosome: my drawing was more correct than my reasoning.

24. METAPODIUS TERMINALIS Dall.

Spermatogonic Divisions. — Two pole views of the chromosomes are shown in Plate XI, Figs. 169, 170. Each shows 2 very minute elements which are unequal in size and are the diplosomes (*Di, di*). Then there is one unpaired, constricted element, the monosome (*Mo*). The remainder are 9 pairs of autosomes (*A, a–I, i*).

Growth Period. — Throughout this period there is a dense safraninophilous body of considerable size close to the nuclear membrane (*Mo*, Plate XII, Figs. 171–173); it is ovoid in the synapsis, more elongate in the postsynapsis, ovoid again in the (incomplete) rest stage; it never appears double as if formed by the conjugation of two elements, nor any at any period shows clearly a longitudinal split. This is probably the monosome because it is far too large to be the bivalent diplosome. No sign at all of the diplosomes is to be seen; this may be either on account of their very small size, or perhaps on account of their not retaining a compact form. The 18 autosomes conjugate end to end to form 9 bivalent ones.

First Maturation Division. — In the prophases (Fig. 174) reappear the diplosomes (*Di, di*) as a pair of small rounded bodies, not attached together until the time of disappearance of the nuclear membrane. In the spindle the 11 elements show a very regular disposition (Figs. 176, 177) like that of *Anasa tristis*, with the bivalent diplosome in the center and the monosome (*Mo*) excentric. All these elements are shown on side view in Fig. 175: there the diplosome is seen to have its components of dissimilar volume, and to divide reductionally as do the 9 bivalent autosomes. But the monosome (*Mo*, Fig. 175), when examined in profile, is seen to be placed with its long axis in the plane of the equator and to divide through its length. As the daughter chromosomes separate in the anaphase (Fig. 178) a constriction upon each marks the reopening of the longitudinal split; but the monosome (*Mo*) does not show this constriction, and upon pole views of a daughter plate (Fig. 179) appears simply ovoid while all the others are dumbbell-shaped.

Second Maturation Division. — In the spindle the chromosomes are again differently arranged (Fig. 180), they are 11 in number; the diplosome (*di*) can be recognized by its small size, the monosome (*Mo*) by its small depth. All of these divide again except the monosome which passes without division into one of the spermatids (*Mo*, Figs. 181, 182).

Literature. — In my previous brief account (1901*b*) I did not describe the second maturation division, gave the number of spermatogonic chromosomes as 22 (counting the constricted monosome as 2), and in the growth period confused the monosome with the diplosomes.

LYGÆIDÆ.

25. ŒDANCALA DORSALIS Say.

Spermatogonic Division. —The spindle contains 13 elements (Plate XII, Fig. 183). These are: 2 diplosomes of approximately equal volume, the smallest of all (*Di, di*); 1 monosome (*Mo*), the only unpaired element; and 5 pairs of autosomes (*A, a–E, e*) of which the pairs are to be recognized rather by peculiarities in form than in size.

Growth Period. — Up to the late postsynapsis the allosomes cannot be distinguished from the autosomes, that is, they neither remain dense and compact nor do they continue safraninophilous. It is, accordingly, probable that until then the allosomes undergo changes parallel to those of the autosomes, except, as will appear from the later history, the monosome remains a single element and the diplosomes probably do not conjugate, while the 10 autosomes go to compose 5 longitudinally split bivalent chromosomes. Throughout there is a large plasmosome (*Pl*, Figs. 184, 185), lying usually against the nuclear membrane. The growth period is closed by an almost complete rest stage (Fig. 185), one in which the chromosomal boundaries cannot be well distinguished. Just before this rest stage there becomes visible a safraninophilous double body (*Mo*, Fig. 184) placed almost invariably upon the plasmosome; we shall find that this is the monosome. It reappears first in the form of a pair of rods, each finely granular, which are to be considered the split halves of the monosome because they are of equal length and volume; at this stage the two are more or less curved so that together they bound an oval space. They soon become compacter with smooth surfaces, and appear as two shorter parallel rods (*Mo*, Fig. 185). No trace of the diplosomes is to be seen, *i. e.*, they do not stain differently from the autosomes.

First Maturation Division. — In the early prophases the plasmosome dissolves without a visible remnant. The monosome (*Mo*, Figs. 186, 187) has the form of two short, thick rods, which may be parallel but are more frequently divergent. The autosomes now commence to stain with saffranine (Figs. 186, 187), and they compose 5 bivalent elements in which each univalent component is longitudinally split; this split gradually narrows up to the stage of the metaphase. And now reappear for the first time the diplosomes (*Di, di,* Figs. 186, 187) as two very small elements, each in structure and stain like a miniature univalent autosome; they are not in contact with each other in any part of the prophase, but are more or less widely separated; sometimes each appears longitudinally split (Fig. 187). By their size relations there can be no doubt which of these various nuclear structures are the diplosomes and which is the monosome. In the late prophases (Fig. 188) the monosome (*Mo*) changes form so that each of its halves becomes spherical; the diplosomes (*Di, di*) become compact and shorter, and though they are usually near together appear never to actually conjugate; and the 5 bivalent autosomes shorten and condense into short tetrads.

In the spindle the diplosomes never form a bivalent element in the equator but always lie on either side and at some distance from this plane (*Di, di*, Fig. 190). A pole view of the equatorial plane shows, accordingly, only 6 chromosomes (Fig. 189), which are the univalent monosome (*Mo*), recognizable by its lesser depth, and 5 autosomes; the constrictions seen on end views of the latter are their longitudinal splits. The monosome is a dyad, while the autosomes are tetrads, as shown on lateral views (Fig. 190). In the anaphase (Fig. 191) each daughter cell receives one of the diplosomes (*Di, di*), a half of the monosome (*Mo*), while the 5 autosomes divide reductionally and their daughter components as they separate show each the reopening longitudinal split.

Second Maturation Mitosis. — Pole views (Fig. 192) of the spindle show 7 elements all in one plane; the smallest is a diplosome (*Di*) while the monosome (*Mo*) may be distinguished from the autosomes by its lesser depth; a lateral view of the same stage is given in Fig. 193, where the monosome is readily marked by its unconstricted form. Each of the autosomes divides equationally and so does the diplosome. But the monosome passes without dividing into one of the spermatids (*Mo*, Fig. 194). A pole view of any spermatid shows a circle of 5 autosomes around a minute central diplosome (Fig. 195): and half of the spermatids show just beneath this chromosomal plate a monosome.

Literature. — I had described (1901*b*) this spermatogenesis in the main correctly, only I failed to decide whether what I called the "odd chromosome" divided in the second maturation division and failed to notice that it is the larger allosome of the growth period; but later (1905) I showed that the monosome does not divide in this mitosis.

26. ONCOPELTUS FASCIATUS Dall.

My preceding account, a rather detailed one, of the spermatogenesis of this species was entirely correct. Of the 16 chromosomes of the spermatogonia I demonstrated that 2 are diplosomes, that these are distinguishable during the growth period, and very frequently separated from each other there, and that they enter the chromosomal plate of the first maturation mitosis separately and that each divides by itself. All that is to be corrected is my former interpretation that each of these is in the spermatogonium already bivalent, and that the division of each in the spermatocytes is to be considered reductional; now I find no good reason for such a view, and judge the latter division to be an equational one of the diplosomes. There is to be added to that former account the description of the

Second Maturation Division. — A pole view of a daughter chromosomal plate of the first maturation mitosis (Plate XII, Fig. 196) shows 9 elements ; the 2 central rounded ones are the univalent diplosomes, and outside of them is a circle of 7 univalent diplosomes the constriction of each being its longitudinal split. As these come to arrange themselves in the equator of the second spindle there appear to be only 8 instead of 9 of them ; this is because the univalent diplosomes have conjugated in the centre to form a bivalent one (Fig. 197). This bivalent element can be recognized only by its central position because its components are of equal volume (*Di, di,* Fig. 198). Each of the 7 autosomes divides equationally, but the bivalent diplosome divides reductionally. And each spermatid exhibits always exactly 8 elements of which the central one is a diplosome (Fig. 199).

27. PELIOPELTA ABBREVIATA Uhler.

Spermatogonic Division. — There were on my preparations only two fairly clear pole views of the equatorial plate (Plate XII, Figs. 200, 201), and in each of these the elements were more or less obliquely placed. There are in all 14 chromosomes, 10 of which are noticeably larger and 4 considerably smaller. The following history shows that these 4 smaller ones are diplosomes, which compose a larger pair (*Di. 2, di. 2*) and a smaller pair (*Di. 1, di. 1*).

Growth Period. — From the synapsis stage (Fig. 202) there are in each nucleus, besides the long loops of the bivalent autosomes, 2 large dense bodies of equal volume ; and when the autosomes become longitudinally split each of these becomes constricted at its middle point (*Di. 2, di. 2,* Fig. 203). By their size relations these are evidently the same as the pair of larger diplosomes of the spermatogonia, for they are much too large to correspond to the smaller pair. They may be apposed (Fig. 202) or may be separated (Fig. 203). The smaller diplosomes could not be distinguished with certainty at this time, whence it is likely that they undergo changes like the autosomes do, or at least do not remain dense and safraninophilous. The 10 large autosomes join end to end to form bivalent elements, and each becomes longitudinally split ; they are then mostly in the form of a U or a V and the split in the arm of each remains narrow and never opens up widely.

First Maturation Division. — In the prophases condense 5 large tetrads, which are the bivalent autosomes ; a single one of them is drawn in Fig. 204, and 4 in Fig. 205, they being the bodies that are not lettered ; these may condense so as to appear nearly solid and very massive, but frequently the point of junction of the univalent elements continues recognizable as well as the longitudinal split in each of the latter, and this split is always parallel to the long axis. Next in size to these are 2 elements

(*Di. 2, di. 2*) alike in volume, each transversely constricted and the two never in close contact; each of these is then a dyad, not a tetrad, therefore is univalent and the two correspond to the larger pair of diplosomes of the earlier stages. Then there become clearly distinguishable a pair of much smaller bodies (*Di. 1, di. 1*, Figs. 204, 205) which correspond to the smallest chromosomes of the spermatogonium, and are a smaller pair of diplosomes; in the earlier prophases (Fig. 204) each of them is longitudinally split, and they may or may not be in mutual contact. Therefore there are in the prophases: 5 bivalent autosomes, 2 larger univalent diplosomes, and 2 smaller univalent diplosomes, 9 bodies in all.

In the equator of the spindle there may be the same number of elements, or there may be only 8 (Figs. 205, 206). This results because the smallest diplosomes may be joined end to end (as in Figs. 206, 207, *Di. 1, di. 1*) or be placed side by side (Fig. 208, *Di. 1, di. 1*); in either case, however, a whole one of these passes without division into one of the daughter cells, which amounts to a reduction division of the pair, and to each appear to be attached mantle fibres from only one spindle pole. The 2 larger diplosomes (*Di. 2, di. 2*, Figs. 206–208), which are recognizable by being dyads of equal volume and next in order of size, remain separated from each other, and each by dividing along the plane of its previous constriction divides equationally. The remaining, largest, chromosomes are all tetrads (the unlettered ones of Figs. 206–208), and these divide reductionally, because each divides transversely to its long axis. Each second spermatocyte receives accordingly 5 whole autosomes, a whole diplosome of the smaller pair, and a half of each larger diplosome, a total of 8 elements.

Second Maturation Division. — Here there are on pole views (Fig. 209) always only 7 chromosomes visible, 5 larger and two much smaller. The five largest are clearly the autosomes. The two smaller must then correspond to the 3 diplosomes that each second spermatocyte receives, *i. e.*, one of them must be bivalent. Lateral views (Fig. 210, which shows all the elements) demonstrate that each of the smaller elements is composed of two parts of equal volumes. Therefore there could not have taken place a conjugation of a large with a small diplosome, but two diplosomes of equal volumes must have conjugated. Now since we found that the second spermatocyte receives only one diplosome of the smaller pair, but a half of each of the larger, and since the latter were of equal volume, it is these larger ones that must conjugate, come to lie the one immediately above the other, in the second spindle Accordingly, of the 6 elements shown in Figs. 209 and 210, the 5 largest are univalent autosomes, the smallest (*di. 1*) is one univalent diplosome of the smaller pair, while the next smallest, the central one, is bivalent (*Di. 2, di. 2*). This explanation suffices to make clear the change in number from 8 to 7 in conjunction with the persisting size relations.

Stages later than that of Fig. 210 were not found; but from the form and position of the chromosomes there it is probable that the 5 autosomes divide equationally, that the small diplosome (*Di. 1*) divides in the same way, but that the bivalent diplosome (*Di. 2, di. 2*) divides reductionally.

Accordingly, there are two pairs of diplosomes; in the maturation mitoses the larger of them divide first equationally then reductionally, the smaller first reductionally then equationally, so that the phenomena of division are reversed in the two pairs.

Literature. — In my preceding account (1901c) the spermatogonial number of chromosomes was erroneously given as 16, since I had counted two of the larger constricted ones as two each; and the contrasted behavior of the two diplosome pairs was overlooked because the second maturation mitosis was not studied.

28. Ichnodemus falicus Say.

Spermatogonic Division. — On the clearest pole view (Plate XII, Fig. 211) 15 elements could be counted. There must, however, be 16 present at this stage as will be shown by the later ones. Further, 4 must be diplosomes, of which the two marked *Di. 2, di. 2* must be the larger pair of diplosomes and *Di. 1* be one component of a smaller pair. The 12 largest bodies are certainly autosomes.

Growth Period. — Six bivalent autosomes are found in the form of V's or, as frequently, parallel rods, that is, they may conjugate end to end or side to side; each becomes longitudinally split. Sharply distinguishable from these during the whole growth period are 2 deep-staining, compact bodies, markedly different in volume, attached to the nuclear wall (*Di. 2, di. 2*, Figs. 212–214). These are the larger pair of diplosomes and represent the two similarly lettered ones in the spermatogonium (Fig. 211). They are rarely in contact with each other so that it may be that they do not conjugate. The larger of them (*di. 2*, Fig. 214) becomes longitudinally split, this split continuing up to the following mitosis; the smaller one is elongate, but only in rare cases does it show signs of division (*Di. 2*, Fig. 213). Towards the close of the growth period, which is not a rest stage, a large irregular plasmosome is developed (*Pl.*, Fig. 214), to which one or the other of the large diplosomes is frequently attached.

First Maturation Division. — In the early prophases reappear the pair of small diplosomes (*Di. 1, di. 1*, Fig. 215); they are not connected and each is at first a small bent rod with uneven contours and a longitudinal split. Each condenses and shortens, the split still maintained (*Di. 1, di. 1*, Figs. 216–219), and they usually do not conjugate until the stage of the equatorial plate. The pair of larger diplosomes are recognizable by their greater size (*Di. 2, di. 2*). Then there are in each nucleus 6 bivalent autosomes (Figs. 215–219, all of them shown in Fig. 217), which are much larger than any of the 4 diplosomes; they are at first of very diverse forms, inasmuch as each may have its univalent components meeting at an angle, or placed side by side, or more or less twisted around each other; the longitudinal split may be narrow for its whole length, or may be widest at the middle. These generally condense so that in each the univalent components come to lie in one line and the longitudinal split becomes obscured (Fig. 219).

On pole views of the monaster stage (Figs. 221, 222) are seen always 9 elements. The 6 largest are the bivalent autosomes (those that are not lettered), the smallest one, which is usually central in position, is bivalent being the pair of small diplosomes (*Di. 1, di. 1*) the components of which may lie one above the other or else side by side. The 2 remaining elements are those marked *Di. 2, di. 2*; they are unequal in volume and are placed apart from each other upon the periphery of the chromosomal plate; these are the elements of the larger diplosome pair, each of them univalent. A lateral view of the spindle (Fig. 220) shows the small bivalent diplosome (*Di. 1, di. 1*), the separated univalent diplosomes of the larger pair (*Di. 2, di. 2*), and 3 of the 6 autosomes. The 6 autosomes and the small bivalent diplosome divide reductionally as can be told from their position within the spindle; but each large diplosome by dividing separately undergoes an equation division; each second spermatocyte receives, accordingly, 6 univalent autosomes, one whole univalent component of the smaller diplosome pair, and a half of each component of the larger diplosome pair.

Second Maturation Division. — Pole views of the equatorial plate (Fig. 224) show only 8 elements, and not 9 as in the preceding mitosis. The six largest are the autosomes, and the very smallest is clearly the small diplosome (*Di. 1*). The element lettered *di. 2* must therefore be composed of two elements, in order to account for the apparent reduction in number in the second spermatocyte; and it is indeed bivalent, the composite of the components of the larger diplosome pair, for on lateral aspect of the spindle (Fig. 223) this chromosome is found to be composed of 2 bodies of dissimilar volumes placed end to end (*Di. 2, di. 2*), and we found that the diplosomes of the larger pair were characterized by this dissimilarity in volume. From the position of all these elements in the spindle it becomes evident that all the autosomes divide again, so equationally, and that the small diplosome (*Di. 1*) does the same; but that the bivalent larger diplosome divides reductionally in that its larger component passes into one spermatid and its smaller one into another. Only one good pole view of a spermatid was found (Fig. 225); this showed 7 elements which from their size are to be considered the 6 autosomes and the smaller component of the larger diplosome pair, while the element of the smaller diplosome pair was not visible (though it must be present on account of its division fore-shadowed in the case shown in Fig. 223).

Literature. — In my preceding account (1901*b*) I did not find the diplosomes in the spermatogonic monaster, and did not describe the second maturation division; but I was correct in concluding that there are one bivalent and two univalent diplosomes in the first maturation monaster.

29. CYMUS ANGUSTATUS Stal.

My preparations showed neither spermatogonic mitoses nor pole views of the first maturation division, and their staining was unsuitable for determining the phenomena of the growth period.

Second Maturation Division. — Pole views show 14 elements, one of them (*di. 1,* Fig. 226, Plate XII), very minute and probably a univalent diplosome. Lateral views of the spindle demonstrate that one of the larger elements is composed of two bodies of unequal size placed end to end (*Di. 2, di. 2,* Fig. 228); in one case these two lay side by side (Fig. 227), and each seemed to be connected with only one spindle fibre. This is probably a bivalent diplosome destined to undergo a reductional division. The 13 other elements would seem to divide equationally or at least into equal parts.

While not much can be definitely decided from this stage alone, yet the phenomena show similarity to those of *Peliopelta* and *Ichnodemus*. That is, in the first spermatocyte there might well be 15 elements, one more than in the second; and these would be 12 autosomes that divide reductionally, a small bivalent diplosome dividing in the same manner, and a larger pair of diplosomes each component of which would divide by itself and these two then conjugate in the daughter cell. In the second spermatocyte there is certainly one bivalent element that divides reductionally, and it shows close resemblance to the bivalent diplosome of the same stage in *Ichnodemus*.

Literature. — My preceding observations (1901*b*) stated nothing definite. My preparations of *Cymus luridus,* of which a brief description was given by me (1901*a*), were not favorable for study.

TINGITIDÆ.

30. TINGIS CLAVATA Stal.

No spermatogonic divisions were seen.

Growth Period. — The iron-hæmatoxylin stain of the slides was too deep for clearly distinguishing allosomes, but, in addition to a large, somewhat irregular body that is probably a plasmosome, may be found one or two dense bodies of different volumes that may be diplosomes.

First Maturation Division. — Pole views show in most cases 7 elements (Plate XIII, Fig. 229), a circle of 6 around a central one. On side view all of these appear dumb-bell-shaped (Fig. 230) except the central one which is composed of parts of unequal volumes (*Di, di*); these parts are placed usually end to end but sometimes side by side. This central one is probably a bivalent diplosome and divides reductionally, while the 6 others are probably bivalent autosomes that also divide. In two pole views out of a considerable number seen 8 elements were found; this happens because sometimes the components of one of the autosomes may be separated, as the two bodies marked *M* in Fig. 231.

Second Maturation Division. — There are regularly 7 elements present, namely, 6 autosomes and either the larger (*di*, Fig. 232) or the smaller diplosome (*Di*, Fig. 233). In a single case, manifestly an abnormality, 8 elements were present, both diplosomes being in the same cell (*Di, di*, Fig. 234). All 7 elements divide, presumably equationally, and 7 elements are always present in the spermatids (Fig. 235), half of the spermatids containing a division product of the larger and half of them a division product of the smaller spermatid.

Literature. — In my earlier description (1901*a*) I noted that one of the chromosomes of the first maturation mitosis is characterized "in having its two components of very unequal volume," but I failed to follow its behavior in this and the following mitosis.

PHYMATIDÆ.

31. PHYMATA sp. (*P. wolffii* Stal.?).

I can add little to my former account (1901*b*), and find that the chromosomes are too crowded in the second spermatocytes to be counted with precision. But in the spermatogonium I now think there are 29 and not 30 elements as I had previously described, for one is much longer than any of the others (*Mo*, Fig. 237, Plate XIII), and this I had originally counted as two. This unique chromosome was to be seen in all three of the distinct pole views. Therefore there is a possibility that a monosome is present in this species.

REDUVIIDÆ.

32. ACHOLLA MULTISPINOSA de G.

Spermatogonic Division. — Pole views show exactly 32 chromosomes (Plate XIII, Fig. 238), of which 8 are 4 minute pairs of diplosomes.

Growth Period. — The 4 pairs of diplosomes can be recognized throughout the growth period, and were described in some detail in my previous paper; they lie on the surface of the plasmosome (*Pl*, Fig. 239), and as in the spermatogonium the pairs are of slightly different sizes.

First Maturation Division. — The bivalent diplosomes, 4 in number, are readily distinguished by their small size and lie always upon the periphery of the chromosomal plate ; most frequently 3 lie close together, the 4th some distance off from them (Fig. 241) ; or they may all be near each other (Fig. 242), or 2 may be situated at one place and 2 at another. These diplosomes with the 12 bivalent autosomes are all illustrated on lateral aspect in Fig. 240, and all these elements divide, probably reductionally.

Second Maturation Division — Pole views of the spindle show again 16 elements but in different arrangement in that the 4 diplosomes now lie in the center (Figs. 243, 244). Lateral views show that all of these are bipartite, and therefore they all probably divide again though their number could not be counted in the spermatids. There is certainly no conjugation of any of the diplosomes in the second spermatocytes, and no evidence at any stage of the presence of a monosome.

Literature. — My earlier observations (1901 *b*) were entirely correct, and I have to add to them simply the account of the second spermatocytes.

33. SINEA DIADEMA Fabr.

My earlier observations were essentially correct, and the three pairs of diplosomes of the rest stage of the spermatocyte are shown in Plate XIII, Fig. 245, attached to the plasmosome (*Pl*). Another pole view of a first maturation monaster is presented in Fig. 246, the 3 bivalent diplosomes readily distinguishable by their small volumes. Of the 13 autosomes three are always close together and so form a regular complex (*A*, *a*, *B*, *b*, *C*, *c*), just as I previously described ; but now I find no reason to consider the central one of this complex quadrivalent, for there is no good evidence that it is anything else than an unusually large bivalent autosome and it does not behave differently from the others during the preceding growth period. This central one of the complex is always the largest and a very evident tetrad (*B*, *b*, Figs. 247, 248) ; close to one end of it is a smaller bivalent autosome (*A*, *a*), and close to its other end a still smaller one (*C*, *c*) ; these size relations are always the same. All the elements of this mitosis are shown on lateral view in Fig. 247 ; the 3 smallest are the bivalent diplosomes and they are of slightly different volumes. All 16 elements divide reductionally, so that each second spermatocyte receives a univalent component of each. The complex of the 3 autosomes *A*, *a*, and *B*, *b*, and *C*, *c* divides more tardily than the others, as shown by the successive stages of Figs. 248–250, and in these anaphases the lateral autosomes (*A*, *a* and *C*, *c*) become separated from the large middle one (*B*, *b*).

There were no clear cases of second maturation mitoses. But judging from the composition and behavior of the elements in the first spermatocytes, there would be in the spermatogonium : 6 univalent diplosomes and 26 univalent autosomes.

34. PRIONIDUS CRISTATUS Linn.

My former account (1901*b*) was correct in the main.

A new drawing of a spermatogonic monaster is given (Plate XIII, Fig. 251). Of the 26 chromosomes 2 are much larger (*A*, *a*) and 2 much smaller (*L*, *l*) than the others. All these are found on careful inspection to be arrangeable into a series of pairs, *A*, *a–M*, *m*, in which the two components of each pair are of approximately equal volume except the 2 marked *K*, *k*. There is probably no monosome because the number is an equal one.

In the complete rest stage of the spermatocytes are found 3 or 4 safraninophilous bodies (Fig. 252, *Di. 1*, *Di. 2*, *Di. 3*) attached to the surface of a large, more or less central, plasmosome (*Pl*). They are of unequal volumes ; and when there are 3 of them each appears bipartite, while when there are 4 the 2 smallest are each unipartite. Perhaps, as in *Sinea*, these relations are to be interpreted as 3 bivalent diplosomes, the smallest of which may sometimes have its parts separated.

BELOSTOMATIDÆ.

35. ZAITHA sp.

Spermatogonic Division. — In all of eight clear pole views 24 chromosomes were counted (Plate XIII, Fig. 253). They are of very different volumes, 4 being much larger and 2 much smaller than any of the others. They make up 11 pairs gradated both in form and size (*A*, *a–K*, *k*), all these being autosomes ; and 1 pair of 2 unequal components (*Di*, *di*) that correspond to the diplosomes of the later stages. The 4 largest autosomes are about equal in length, but 2 of them (*A*, *a*) are thicker than the others (*B*, *b*). The 2 smallest elements (*K*, *k*), are always slightly different in volume.

Growth Period. — This terminates with a complete rest stage of short duration. In it is found a single spherical plasmosome (*Pl*, Fig. 254), and attached to its surface either 2 or 3 smaller rounded bodies, *Di. 1*, *di. 1*. The most frequent condition is that figured, and these smaller bodies probably represent the unequal diplosomes of the spermatogonium, the bipartite nature of the larger being due to a splitting. The amount of cytoplasm is relatively great and it contains towards the end of the growth period, besides one or a few small yolk spherules (*Yk*), 3 or 4 rather dense bodies (*Id*) more or less spherical in form, staining like the cytoplasm ; they are variable in position and size but are usually close to the nucleus. Each one has a considerable resemblance in form and size to the single idiozome body of *Peripatus*; and they are probably masses of idiozome substance, well defined and few in number, whereas in most of the Hemiptera this substance is usually more or less diffused in a zone concentric to the nucleus. In the synapsis stage there is a single large mass of this substance at the distal pole of the nucleus.

First Maturation Division. — There are always 13 elements (Fig. 256), one more than half the number in the spermatogonium, therefore 2 of them must be univalent and the others bivalent. They show rather a dense grouping. The largest 2 (*A, a–B, b*) correspond to the 2 largest pairs of the spermatogonium, and are usually placed in the middle of the chromosomal plate; 2 smallest elements always lie on the periphery, the smaller of which (*K, k*) probably represents the smallest pair of the spermatogonium. All divide in this mitosis so that the second spermatocyte receives also 13 chromosomes.

Second Maturation Division. — Here the chromosomes are grouped differently in the spindle (Fig. 258), namely, as a circle of 11 around a central pair. The latter is composed of a smaller (*Di*) and a larger (*di*) body placed one above the other, and these move apart into opposite spermatids before the other chromosomes divide (Fig. 257); these 2 are obviously the unequal elements of the spermatogonia, and each of them must have undergone an equational division in the preceding mitosis and have been univalent there. The smaller component of this bivalent diplosome, *Di*, is next larger than the smallest of the autosomes, *K, k*, while the larger, *di*, is, counting from the smallest, the fourth in size of all the elements; these size relations probably hold true for the preceding division, and by means of it we can determine which elements of the former chromosomal plate (Fig. 256) are these elements *Di* and *di*. Each of the 11 autosomes divides, so that each spermatid receives 12 elements in all; this is to be determined from the form of the chromosomes and their position in the spindle (Fig. 257), for they are too densely crowded in the spermatids to be determined there.

Literature. — My preceding account (1901*b*) was entirely correct, except that by a slip of the pen I stated that the second spermatocyte receives only 11 chromosomes; I did not describe the second maturation mitosis.

HYDROBATIDÆ.

36. HYGOTRECHUS sp.

Spermatogonic Division. — There were only four good pole views. In three of them 20 elements could be counted, but in the fourth, which was the clearest because the chromosomes were most fully separated, 21 were found (Plate XIII, Fig. 259). Twenty of these are seen to form 10 pairs (*A, a–J, j*), which vary to considerable extent in both form and volume; but the very smallest (*Mo*) has no mate in size, and is therefore a monosome.

Growth Period. — This terminates in a complete rest stage (Fig. 260). There is a large plasmosome (*Pl*) attached to which is either a single body or a pair of bodies of like volume (*Mo*); the latter condition is to be explained as a monosome divided equationally into two parts, because these later join to compose the monosome of the maturation mitoses, and more particularly because in the earlier growth period these are represented by a single one. This monosome, respectively its halves, swells considerably in size during the growth period, and while continuing dense it does not remain safraninophilous. No bodies were found that represented diplosomes.

First Maturation Division. — In the prophases the plasmosome disappears; Fig. 261 reproduces a late prophase and shows all the chromosomes. Each autosome is bivalent, composed of 2 univalent ones placed more usually end to end, more rarely side to side, and each univalent element when viewed from its flattened surface shows a split along its axis which is evidently the same as the earlier longitudinal split of the postsynapsis stage. This split gradually closes, though never completely, as the autosomes condense and retains its position parallel to the length of the autosome. Besides these autosomes there are 2 much smaller bodies (*Mo*), which are alike in size and each, so far as I could determine, is unipartite; at this stage they are frequently not separated but apposed, and probably represent the halves of the monosome.

Pole views of the equatorial plate (Figs. 266, 267) show 11 elements, one more than half the number in the spermatogonium; on strict pole view 10 of them, the autosomes, always seem bipartite, while the smallest one, the monosome (*Mo*), appears unipartite; seen from the side (Fig. 262) the 10 autosomes are found to be tetrads, while the monosome (*Mo*) is a dyad. This monosome divides and apparently through the plane where its halves had previously come together, therefore equationally. The 10 tetrads, the bivalent autosomes, are so nearly quadratic in outline that it is difficult to decide how they divide, but there is no reason to hold that they do not divide reductionally. As a result each second spermatocyte receives also 11 elements.

Second Maturation Division. — The chromosomes evince no great constancy in their arrangement in the spindle (Figs. 266, 267), the monosome may be recognized by its lesser depth (*Mo*). Side views (Fig. 265) show that 10 are always bipartite with their constrictions placed in the equator; these are the autosomes and there can be no question that all of them divide. But the smallest element, the monosome (*Mo*), is spherical, and placed usually a little above or below the plane of the autosomes; I have not drawn its mantle fiber attachments because I was unable to ascertain them. Only one clear pole view of a daughter plate of chromosomes of this mitosis was seen

(Fig. 268), and that showed 10 elements. But from its unipartite appearance in the spindle, and from its situation a little out of the plane of the autosomes, there can be little doubt that the monosome passes undivided into one of the spermatids.

Literature. — My former description (1901) was incorrect in concluding 20 to be the normal number of chromosomes, and in supposing the allosomes of the growth period to be a pair of diplosomes. Also I did not describe the second maturation mitosis.

37. LIMNOTRECHUS MARGINATUS Say.

The spermatogenesis is on the whole very similar to that of the preceding species. There were no spermatogonic divisions on my slides.

Growth Period. — There is a monosome, which in the rest stage (*Mo*, Fig. 269, Plate XIII) is longitudinally split; it may be nearly spherical, but more usually is elongate with the split along its length; further, it is usually separated from the plasmosome (*Pl*). These constitute the main differences from *Hygotrechus.*

First Maturation Division. — There are 10 large tetrads, the autosomes, and 1 small dyad, the monosome (*Mo*, Figs. 271, 272). All of them divide, the monosome equationally.

Second Maturation Division. — There are also 10 autosomes and the half of the monosome (Fig. 274), the latter recognizable upon pole view by its lesser depth. All the autosomes divide, but the monosome (*Mo*, Fig. 273) remains rounded, is placed usually a little nearer one spindle pole than the other, and therefore probably passes undivided into one of the spermatids.

Literature. — My preceding account (1901*b*) was very brief, and I supposed a pair of diplosomes to be present.

CAPSIDÆ.

38. CALOCORIS RAPIDUS Say.

Spermatogonic Division. — There was only one clear pole view (Plate XIII, Fig. 275), and that showed exactly 30 elements.

Growth Period. — Throughout this period there is a deep-staining, rod-like body close against the nuclear membrane, which on profile gives the effect of a crescent. In the synapsis (Fig. 276, *Mo. 1*) it is more or less ovoid, but it later assumes the form of a bent rod (*Mo. 1*, Fig. 277) and during all the stages except the earliest shows a well-marked longitudinal split. In the later stages this body has usually the form of two bent rods, which may be parallel, or slightly divergent when the space between them is the longitudinal split. This is the larger monosome of the spermatocytes, as will be demonstrated by its later history. Though always prominent in the nucleus by reason of its large size and deep stain, it does not remain completely compact and dense, but sometimes shows a loosening of its texture. Besides this there is a second and much smaller monosome (*Mo. 2*, Figs. 276, 277), usually rod-shaped in the synapsis and more spherical later, generally separated from the nuclear membrane; it shows no signs of a longitudinal split. Both of these monosomes increase considerably in volume, then decrease again during the following prophases. Plasmosomes seem to be absent, and there is no complete rest stage.

First Maturation Division. — In the prophases (Fig. 278) the smaller monosome (*Mo. 2*) can be recognized by its unipartite aspect, the larger one (*Mo. 1*) by its form of two more or less parallel rods. All the other elements are quadripartite autosomes except the two smallest; one of the latter has the shape of two apposed spherules (*Di. 1*, Fig. 278), while the other (*Di. 2*) eventually assumes this form but is the latest of all the chromosomes to become dense in structure; these two smallest elements are probably bivalent diplosomes, because though they are not distinguishable during the growth period they differ from the monosomes by much smaller volume and different form; and I judge that each is bivalent on account of its behavior in the two maturation mitoses.

In the spindle there are always 16 elements, all placed in one plane except one (*Mo. 2*, Figs. 279–283) that lies invariably nearer one spindle pole than the other. This is the only one that seems unipartite, and is the smallest of all; it is undoubtedly the smaller monosome, and has decreased in volume since the prophases. Of the remaining elements one is the larger monosome and it can be recognized on side view only, and then because its long axis lies in the plane of the equator (*Mo. 1*, Fig. 283). Then there are 2 diplosomes (*Di. 1, di. 2*) which are very small and next larger than the smaller monosome. The 12 remaining elements are 12 bivalent autosomes, each quadripartite; one of them, that marked *t* in the Figs. 279–281, is unusually large, and for this reason I had originally (1901*b*) supposed it to be quadrivalent; but since there are 30 elements in the spermatogonium this one cannot be more than bivalent.

The 12 bivalent autosomes divide transversely to their lengths, therefore probably reductionally. The two diplosomes also divide, but in what way I have no means of determining. The larger monosome divides and equationally. But the smaller monosome, which always lies a little out of the plane of the other elements, never divides but passes wholly over into that spermatocyte of the second order to which it is nearest. Half the second spermatocytes receive, accordingly, 16 chromosomes, and half of them 15, the one that may be lacking being the smaller monosome.

Second Maturation Division. — Pole views of the second spindle are shown in Figs. 285, 286. One of them is a cell containing the smaller monosome (*Mo. 2*, Fig.

285), while the other is a cell that lacks this body. There are always two diplosomes that can be recognized by their small size, but slightly larger than the smaller monosome. As in the preceding mitosis the smaller monosome always lies a little outside of the plane of the other chromosomes, so in this second mitosis the larger one always lies somewhat to one side of the equatorial plane (*Mo. 1*, Fig. 284) ; and by virtue of this position it may be recognized even upon pole view (*Mo. 1*, Fig. 285). Fig. 284 shows the 3 smallest elements, which we have found to be the smaller monosome (*Mo. 2*), and the two diplosomes (*Di. 1, Di. 2*), all three of them showing a division constriction. This demonstrates that the smaller monosome divides, that the diplosomes also do, and because the 12 autosomes are equally constricted they too must divide. But the larger monosome (*Mo. 1*, Fig. 284) lies nearer one spindle pole than the other, is never constricted, and in the anaphases (Fig. 287) passes without dividing into one of the spermatids.

Accordingly there are in this complicated case : 12 autosomes that divide in both mitoses, 2 diplosomes that do likewise (therefore are probably also bivalent), a smaller monosome that does not divide in the first but does divide in the second mitosis, and a larger monosome that divides in the first but not in the second mitosis. Therefore, each spermatid receives 12 autosomes and 2 diplosomes, while only half of them receive the larger, and only half of them the smaller diplosome; whether any spermatid ever receives both monosomes, or whether any one ever lacks both monosomes, I could not decide, because the chromosomes are closely crowded in the spermatids.

From the relations of the chromosomes in the spermatocytes the elements in the spermatogonium should be as follows : 24 autosomes, 1 larger and 1 smaller monosome and 4 diplosomes, a total of 30 elements which was the number constated to be present there.

Literature. — In my earlier observations (1901*b*) I mistook the larger monosome of the growth period for a plasmosome, because I supposed a plasmosome must be present ; what I then called the "univalent chromatin nucleolus" corresponds to what I now denominate the smaller monosome ; and I correctly showed that this does not divide in the first· maturation mitosis. The following mitosis was not described. Otherwise the complex phenomena were correctly ascertained.

39. PŒCILOCAPSUS GONIPHORUS Say.

Growth Period. — This is terminated by a complete rest stage. Attached to the plasmosomes (*Pl*, Fig. 288, Plate XIII), though occasionally separated from them, are a number of safraninophilous dense allosomes. The largest of these (*di. 1*) is always in the form of a pair of short parallel rods, and, therefore, is to be regarded as probably a longitudinally split, univalent element. Three other pairs of different sizes are always to be seen (*Di. 1, Di. 2, Di. 3*) and sometimes a fourth (*Di. 4*). The components of each pair are equal in volume, but whether each pair is to be considered as two diplosomes, or as the division products of a single one, I could not determine since the number of chromosomes in the spermatogonia is unknown.

First Maturation Division. — There are always 18 elements (Fig. 289), 17 large and 1 (*Di. 1*) much smaller. The latter is always bipartite (Fig. 290), never quadripartite, and as will be evident from its later history is an univalent diplosome, and from its size perhaps correspondent to the two bodies marked *Di. 1* in the growth period (Fig. 288). Of the 17 larger elements 1 must be the largest diplosome of the preceding growth period (*di. 1*, Fig. 288), but at this stage it cannot be distinguished with certainty from the other larger elements. In this mitosis the other small diplosomes of the growth period (*Di. 2, Di. 3, Di. 4*) are to be found neither in the spindle nor in the cytoplasm. All 18 elements divide, and this is an equation division of the large and small diplosome, but probably a reduction division of the 16 bivalent autosomes.

Second Maturation Division. — There are 17 larger elements seen on pole views (Fig. 291), 1 less than in the preceding spindle. This is because the large and small diplosome have conjugated end to end, as one may ascertain by careful focussing (*Di. 1, di. 1*). Lateral views (Fig. 292) show that this bivalent element lies always slightly out of the plane of the other chromosomes, and that each component of it is unconstricted. Each of the 16 autosomes divides, but the components of the bivalent diplosome pass without division into opposite spermatids. Two daughter plates of the anaphase are reproduced, as drawn from the same cell at two levels; one exhibits the smaller diplosome (*Di. 1*, Fig. 293), while the other lacks this but shows the larger diplosome (*di. 1*, Fig. 294).

From the number of chromosomes in the maturation mitoses it may be concluded that there are present in the spermatogonia 32 autosomes and 2 diplosomes.

Literature. — My previous account (1901*b*) confused the two maturation mitoses, and did not describe the second one.

40. Lygus pratensis Linn.

Spermatogonic Division. — There were only 2 pole views, on the one I counted 33, on the other 34 elements. The correct number is probably 35 as we shall find.

Growth Period. — One large, longitudinally-split allosome can be distinguished in the spermatocytes; whether there are others could not be determined.

First Maturation Division. — In the spindle there are 19 chromosomes (Plate XIII, Figs. 295, 296). The smallest of them (*Mo*, Fig. 296) is never in the equatorial plane but always nearer one of the spindle poles; it does not divide but passes bodily into one of the spermatocytes of the second order. This minute element would appear to be a monosome, comparable to the smaller monosome of *Calocoris*. There is no sign of it in the chromosomal plate of the following mitosis. Of the 18 elements that lie in the equator (Fig. 295) all divide in this mitosis. Two of them (*Di. 1* and *Di. 2*, *di. 2*) are much smaller than the others; the smaller of the two (*Di. 1*) is a univalent diplosome as its later behavior shows, while the larger is a bivalent element and it may be a pair of diplosomes (though its small size is the only reason to consider it a diplosome). Of the 16 large elements one of the largest, if not the very largest, must be another univalent diplosome, which with the small element *Di. 1* are unequal components of a diplosome pair.

Second Maturation Division. — There are always exactly 17 elements to be seen on pole views of the spindle (Fig. 297), 2 less than in the preceding spindle; this number was found in numerous cases. All are larger than the small monosome of the antecedent mitosis, and this monosome is not to be found in the chromosomal plate; one would expect to find it in the equator of half of the second spermatocytes, as is the case with the correspondent element in *Calocoris;* but it is always absent, and therefore probably lies out in the cytoplasm where it is indistinguishable from small yolk spherules. Further, in the equator there is only one separate small element (Fig. 297, *Di. 2*), and not 2 separate elements (as in the preceding spindle, Fig. 295, *Di. 1*, *Di. 2*). Careful study shows that one of the chromosomes is bivalent, composed of a small one (*Di. 1*, Fig. 298) placed at the end of a much larger one (*di. 1*), the larger one lying invariably a little above or below the equator which enables one to recognize it upon pole view (*di. 1*, Fig. 297). This bivalent chromosome is composed of the division products of the largest and smallest diplosomes of the first spermatocytes, which had divided separately but are now in conjugation. The single separate small element (*Di. 2*, Figs. 297, 298) again divides by itself; it is a little larger than the smaller element of the bivalent pair and therefore represents a half of the bivalent element *Di. 2, di. 2* of the former mitosis. The 15 autosomes also divide, and the bivalent diplosome divides reductionally, its smaller component going into one spermatid and its larger one into the other; for this becomes evident from their position within the spindle (Fig. 298, *Di. 1, di. 1*), while in the anaphases the larger component (Fig. 299, *di. 1*) comes to lie wholly in one of the daughter chromosomal plates.

There are accordingly in the maturation mitoses: one very small monosome that does not divide in the first spermatocyte, and is not present in the chromosomal plate of the second ; a large and small diplosome (*di. 1, Di. 1*) that divide separately and therefore equationally in the first mitosis, but conjugate in the second spermatocytes and undergo a reductional separation there ; and a small bivalent element, *Di. 2, di. 2*, that may be another diplosome, which divides in both mitoses as do the 15 autosomes. Consequently each spermatid must receive halves of the 15 autosomes and of the element *di. 2, di. 2*, half of them receive *Di. 1* and the other half receive *di. 1*, and half of them get the monosome.

From these relations we may conclude for the spermatogonium : 30 autosomes, one monosome, one large and one small diplosome (*di. 1, Di. 1*), and a pair of small diplosomes (*Di. 2, di. 2*), a total of 35 elements.

Literature. — In my earlier account I overlooked the small monosome, and did not describe the second maturation division.

II. GENERAL CONSIDERATIONS.

1. BEHAVIOR AND SIGNIFICANCE OF THE ALLOSOMES.

In the Hemiptera heteroptera the allosomes present the following relations in the spermatogenesis :

A. *Only Diplosomes Present*, and these exhibiting the following differences :

A1. The diplosomes conjugate early in the growth period, divide reductionally in the first maturation mitosis, and equationally in the second. This is the case in *Tingis*, where there is a single pair with components of very unequal volume ; and in *Acholla* (4 pairs) and *Sinea* (3 pairs), where the diplosomes are very small and the components of a pair of about equal volume. In *Sinea* and *Acholla* they remain dense during the growth period ; in *Tingis* it was not determined how they behave during this stage.

A2. One pair of diplosomes which divide separately and equationally in the first maturation mitosis, but in the second spermatocytes conjugate and then divide reductionally. This modus was first discovered by Wilson ; I had shown (1901*b*) that in certain species (*Euschistus tristigmus, Oncopeltus, Zaitha*) the diplosomes divide separately in the first maturation mitosis, but I failed to note, because in these species I omitted to describe the second mitosis, that their daughter products unite in the second spermatocytes and there undergo a reductional division. Diplosomes of ·this behavior Wilson called the "idiochromosomes," and he correctly noted that they are unequal in volume ; in *Nezara* alone he states that they are equal, but even here I find that there is always a slight voluminal difference. They always remain more or less dense and compact during the growth period ; and in most cases they conjugate early in the growth period as I had previously described, but, as Wilson first demonstrated in detail, separate from each other before taking position in the first maturation spindle. Wilson has described these for *Lygæus, Cœnus, Nezara, Euschistus, Brochymena, Podisus, Trichopepla;* and they are described in the present paper for *Euschistus, Podisus, Mormidea, Cosmopepla, Nezara, Brochymena, Perillus, Cœnus, Trichopepla, Eurygaster, Peribalus, Oncopeltus, Zaitha,* and *Pœcilocapsus.* In the last named species and in *Trichopepla* much more minute allosomes are found in the growth period, but cannot be distinguished with certainty during the maturation mitoses.

A3. Two or more pairs of diplosomes of diverse behavior. In *Nabis* there are in the spermatocytes two bivalent diplosomes that remain compact during the growth period, divide reductionally in the first maturation division and equationally in the second, and the components of a pair are equal in size ; and then another pair of diplosomes that are of very unequal size, which are also distinct during the growth period, but which divide separately and equationally in the first maturation mitosis and in the next mitosis (without conjugation in the equatorial plate) divide reductionally. In *Peliopelta, Ichnodemus* and probably *Cymus* there is a smaller pair, which do not remain compact during the growth period and do not conjugate until late, and these divide reductionally in the first maturation mitosis and equationally in the second ; and besides these there is a larger pair of very unequal components which remain apart from one another during the growth period and then retain their dense structure, which divide separately and equationally in the first maturation mitosis, and in the second spermatocytes conjugate in the equatorial plane and then divide reductionally. Then in *Syromastes* Gross has described two pairs of diplosomes : the larger conjugate very early in the growth period, remain dense, divide in the first maturation mitosis reductionally and in the second equationally ; while the smaller pair, adequal in volume, undergo changes like the autosomes during the growth period, do not conjugate until after it, and compose a tetrad which divides in the first maturation mitosis but not in the second. Accordingly, this third type of diplosome relations may be said to be a combination of the previous two.

B. *Only Monosomes Present.* — This would appear to be the most unusual condition present in the Hemiptera, and is here described for *Hygotrechus* and *Limnotrechus,* while Henking found it for *Pyrrhocoris;* in these cases the monosome remains compact during the growth period, divides equationally in the first maturation mitosis and does not divide in the second.

C. *Both Diplosomes and Monosomes Present,* showing the following diversities :

C1. One pair of diplosomes of small and adequal volume that usually conjugate in the early growth period and during it may either remain compact or may undergo changes much like those of the autosomes (*Alydus, Metapodius*), divide in the first maturation mitosis reductionally and in the second equationally ; and one monosome, much larger than the bivalent diplosome, always compact in the growth period (except in *Œdancala*, and in *Harmostes* it may become more or less reticular), which divides equationally in the first maturation mitosis, but does not divide in the second. This condition was first described by me 'for *Protenor* and *Œdancala*, then found by Wilson for *Anasa*, *Alydus* and *Harmostes*, and in the present paper it is described for these genera as well as for *Corizus*, *Chariesterus* and *Metapodius*. Accordingly, *Syromastes* would appear to be the only Coreid thus far described which does not conform to this type.

C2. In *Calocoris* there are two bivalent diplosomes that divide in the maturation mitoses first reductionally and then equationally ; a smaller monosome that does not divide in the first maturation mitosis, but does divide in the second ; and a larger monosome that divides in the reverse order of this. The monosomes remain compact during the growth period, but the diplosomes do not.

C3. In *Lygus* there is a single, very small monosome that does not divide in either maturation mitosis. And one pair of diplosomes of very unequal volume, which divide separately and equationally in the first maturation mitosis, conjugate in the second spermatocytes and divide reductionally. Another bivalent element, the smallest, which divides like the autosomes, may be another diplosome pair, but this could not be distinctly determined by me.

C4. In *Archimerus* Wilson (1905c) finds that the monosome does not divide in the first maturation mitosis, but in the second divides equationally ; while a bivalent diplosome with small components of equal volume divides first reductionally and second equationally.

C4. And in *Banasa* Wilson (1905c) describes a monosome that behaves like that of *Archimerus*, together with a pair of very unequal diplosomes that divide in the first maturation mitosis separately and equationally, conjugate in the second spermatocytes, and then divide reductionally.

The other groups where allosomes are known to occur are the following. In the spermatogenesis of the Orthoptera according to the researches of Wilcox (1895), McClung (1899–1905), Sutton (1900, 1902b), de Sinéty (1901), and Baumgartner (1904) there is a single monosome said not to divide in the first maturation mitosis but to divide equationally in the second. The only exceptions among the Orthoptera are *Syrbula*, where I showed (1905) there to be a pair of diplosomes which conjugate early in the growth period, and divide first reductionally and then equationally in the maturation mitoses ; *Hippiscus* as described by McClung (1900), where a single monosome is stated to divide in both maturation divisions ; *Stenopelmatus*, where Miss Stevens

(1905*b*) finds the monosome to disintegrate in the second spermatocyte but to probably reappear in the spermatids; and in *Periplaneta* where Moore and Robinson (1905) conclude there is no allosome, but reinvestigation of this species is needed because Miss Stevens has described a monosome in the closely related *Blattella*.   McGill (1904) has described for *Anax*, an Odonate, an allosome that divides in the first maturation mitosis and not in the second; but this author identifies this single element with a pair of chromosomes of the spermatogonium, which makes the phenomena somewhat difficult to interpret.   The account of the spermatogenesis of the coleopteron *Cybistes*, given by Voinov (1903), I have not seen.   Miss Stevens (1905*b*) finds them to be absent in aphids and *Termopsis* (a termite); in the coleopteron *Tenebrio* she describes a pair of very unequal diplosomes that divide in the maturation mitoses first reductionally and then equationally; and in *Sagitta* she describes an allosome that divides in both maturation divisions.   In *Agalena* Miss Wallace (1905) finds a pair of diplosomes that do not divide in either maturation mitosis, which is quite different from my own results upon *Lycosa* (1905), to the effect that the pair of diplosomes divide reductionally and then equationally.   The spermatogenesis of the Chilopods (*Scolopendra*), as described by Blackman (1905*a, b*). is peculiar in that the monosome during the growth period comes to contain all the autosomes, so to form a "karyosphere"; they pass out of it before the first maturation mitosis, where it does not divide, but it divides equationally in the second mitosis; essentially similar results were obtained by Miss Medes (1905) for *Scutigera*.   Some of the most interesting and complex relations of monosomes have recently been found by McClung (1905) in various acridiids, consisting in the adhesion of the monosome to one or more autosomes whereby plurivalent elements may be formed not only in the spermatocytes but even in the spermatogonia.

We may now attempt to decide what decisions the diversity of behavior of the allosomes, particularly in the Hemiptera, may give in regard to their genesis and mutual relations.

Since Henking's first discovery of them in *Pyrrhocoris* all observers have been in agreement that they are modified chromosomes.   And on the observational basis that we have to-day we are in position to conclude what this genesis may have been. In the first place the ordinary chromosomes, the autosomes, of the Hemiptera are proven to divide in the maturation mitoses first reductionally, and second equationally.   The results of Henking, Paulmier, Stevens and myself are in agreement on this issue, and only Gross assumes a reversed order of division; Gross's position is not borne out by his own observations, as I pointed out in another place (1905) and there reasoned, and Grégoire (1905) has strongly seconded me in this, that probably in all Metazoa the first maturation division is reductional and the second equational.   On account of the great mass of evidence upon this question, which has been fully discussed in earlier papers of mine, we shall assume it as proven that in the Hemiptera the autosomes divide in this sequence. Therefore, the allosomes being modified chromosomes, those allosomes that divide in the same way as the autosomes do would be genetically closest to the autosomes. Such are the diplosomes of the Coreidæ (except the smaller pair of *Syromastes*), of the Reduviidæ and *Tingis, Calocoris*, the smaller diplosomes of *Nabis*, and one of the diplosome pairs of *Peliopelta* and *Ichnodemus*. These diplosomes correspond to the " M-chromosomes " of Wilson. They are in most cases the smallest of all the chromosomes, sometimes very minute, and, except in *Tingis*, are only very slightly different in size. Probably those of them that do not remain dense but become reticular in the growth period, as is the case in *Alydus, Metapodius, Œdancala* and *Calocoris*, are the least modified, because the most similar in behavior to the autosomes. The other kind of diplosomes correspond to what Wilson has called the "idiochromosomes," and he first distinguished between these and the preceding kind. These usually do, sometimes do not, conjugate in the early growth period, enter the chromosomal plate of the first maturation mitosis separately, aud divide there equationally, then in the second spermatocytes (usually but not always after a conjugation in the center of the chromosomal plate) divide reductionally; they always remain more or less dense and compact during the growth period, and are usually very different in volume, though Wilson has shown that in *Nezara* they are nearly equal. Both kinds of diplosomes may occur in the same cell.

We do not know intermediates between these two kinds of diplosomes, though there can well be no doubt that the second is a further modification of the first; because sometimes in the first type the diplosomes may be unequal, and in the second type sometimes almost equal in size, size difference cannot be taken as a criterion of them, and for this reason it seemed to me inadvisable to consider them as quite different allosomes as Wilson has done. The most striking difference between the two types is the discord with regard to the reduction division; in the first type it occurs in the first maturation mitosis, in the second type in the succeeding mitosis. This certainly stands in some relation with the time of conjugation of the elements of the pair, which in the first type is always early in the growth period, while in the second type it may occur then, but frequently does not take place until the stage of the second spermatocyte or may not occur even at that stage. From the series of facts now at hand, we might conclude that the genesis of the diplosomes is as follows. First a pair of autosomes became modified so as to retain their compact nature during the growth period, still maintaining their approximate equivalence in volume. Because such allosomes are usually very small, we might conclude also that they arose from the smallest pair of autosomes. In the next change would appear a growing disparity in size, which, if our last assumption be correct, would be due not to one becoming smaller and to the other becoming larger, but rather to one retaining its original volume and to the other becoming much larger. This second step would then be one of differentiation of the two, a becoming-different, probably implying also difference of metabolic activites. This would account for the lessening affinity of the two as exhibited by the protraction of the time of conjugation. Then would be attained the stage of the second type of diplosomes, no longer united but separate in the first maturation spindle. And the last step would be that, instead of a reduction division of them in this spindle, there would take place there an equational division of each.

In this interpretation, which serves at least to unify the diverse phenomena and is in accord with them, we learn that the two kinds of diplosomes are not really radically different structures, but are rather extremes of a series of modifications.

We may now pass to the question of the genesis of the monosomes. In most cases these are larger than the diplosomes, sometimes the largest of all the chromosomes, more rarely are they very minute, as in *Calocoris* and *Lygus*. Usually the monosome remains dense and compact during the growth period, but in *Œdancala* it becomes reticular and is then practically indistinguishable from the autosomes; in *Harmostes* it becomes reticular to a much less degree. A monosome like that of *Œdancala* is clearly a less modified chromosome than are the monosomes of the other Hemiptera. Then monosomes may divide in the first maturation mitosis but not in the second (*Hygotrechus, Limnotrechus, Pyrrhocoris*, all the Coreidæ except *Syromastes, Œdancala*, and the larger monosome of *Calocoris*); my recent observations show that it is always an equation division, along the line in which the monosome splits in the growth period. But in *Archimerus* and *Banasa*, according to Wilson, the monosome does not divide in the first maturation mitosis but does in the second; I find the smaller monosome of *Calocoris* behaves in the same way, and that in *Lygus* the minute monosome does not divide in either mitosis. Thus with regard to the sequence of division, three kinds of monosomes occur in the Hemiptera, of which the kind that divides reductionally in the first maturation mitosis must be considered the least modified because the one that behaves most like the autosomes.

In an earlier paper (1901*b*) I discussed the question of the genesis of the monosomes; showed that a monosome might be produced by the hybridization of species with different chromosomal numbers, but concluded this to be improbable; and inclined to the view that monosomes arose by some abnormality in mitosis, as by failure of two spermatogonial chromosomes to separate, which led to my assumption that the larger monosomes are bivalent elements. This idea of the bivalence of the monosomes I carried out further in my last paper (1905). This seemed to me to best explain the usually relatively large size of the monosomes. Since then McClung (1905) has demonstrated the occurrence of undoubted bivalent chromosomes in the spermatogonia of certain Orthoptera, which may be a union of two or more autosomes or of a monosome with an autosome.

But Miss Stevens (1905b) showed for *Tenebrio* that while in the spermatogenesis there is a pair of diplosomes of very unequal volume, this pair is represented in the ovogenesis by two of equal volume. Then Wilson (1906b) compared the ovogenesis and spermatogenesis in a series of Hemiptera, confirming Miss Stevens' conclusion and elaborating it; Wilson's results may be briefly summarized as follows. Where there is a single monosome in the spermatogenesis (as in *Protenor, Harmostes, Anasa* and *Alydus*) there are two in the ovogenesis so that the ovogonia possess always an equal number of chromosomes. And where in the spermatogenesis there is a pair of diplosomes of unequal volume, there is in the ovogenesis a pair with components equal in volume to the larger diplosome of the spermatogenesis. Thus while half the spermatids lack the monosome, and half of them lack the larger diplosome, each ovotid would contain a monosome and each a larger diplosome. And from this phenomenon Wilson concludes, as did Miss Stevens before him, that a spermatozoön containing a monosome or the larger diplosome on fertilizing an egg produces a female individual; but that a spermatozoön lacking either of these gives rise to a male individual.

The point in this important discovery of Wilson's that immediately concerns us is that the modification of autosomes into allosomes has taken place in the spermatogenesis; and that a monosome of the spermatogenesis has originated by the continuance of the larger element of a diplosome pair in the sperm cells, and the loss of the smaller element there. This is a very plausible conclusion, but there are in particular two phenomena that must be explained before it can be accepted. One is, how an allosome becomes lost in the spermatogenesis; and the other is, how the allosomes introduced by the spermatozoön into the ovum behave during the ovogenetic cycle; on both of these questions we know as yet practically nothing. I showed in 1904 for *Anasa* that the pair of minute diplosomes of the spermatogonium are represented in the ovogonium by a pair equivalent in size and appearance. Such equivalent diplosomes we have just found to be probably the least modified kind of allosomes. The commencement of the allosomes may have had then a parallel course in the two sexes. And the point that now needs to be determined is the behavior of the ovogenetic allosomes in the growth period and the maturation divisions.

So we have reached the conclusion that the allosomes are to be considered modified chromosomes, of which the most primitive condition would be pairs of like volume conjugating and dividing in the same way as the autosomes do. One component of each pair must be paternal and one maternal, as I proved some years ago (1901b). Therefore, corresponding elements must have become modified in the germ cells of both sexes. A more modified condition would be pairs composed of components of dissimilar volume, not conjugating until the second spermatocyte, and dividing in the maturation mitoses in reverse order from that of the autosomes. Wilson's observations would indicate that this further specialization has taken place in the spermatogenesis alone, but it is by no means proven that such need to have been the case in all species. Finally, as to the monosomes, they may be single surviving components of diplosome pairs of which one has been lost in the spermatogenesis as Wilson concludes; or it is possible that they may have originated by the permanent coalescence of two chromosomes, either autosomes or diplosomes, as I have argued. I wish simply to indicate how diverse the possibilities are, and to point out that we cannot be sure of these conclusions until more is known of the phenomena in the ovogenesis.

As to the function of the allosomes, Paulmier (1899) concluded them to be degenerating chromatin masses: "I would make the suggestion . . . that these small chromosomes, or idants (to adopt for the moment Weismann's terminology) contain "ids" which represent somatic characters which belonged to the species in former times, but which characters are disappearing." Then I argued (1901b): "The chromatin nucleoli [allosomes] are in that sense degenerate, that they no longer behave like the other chromosomes in the rest stages; but they would appear to be specialized for a metabolic function. Thus it might be that in the insects the chromatin nucleoli are those chromosomes which exert a greater metabolic activity than the other chromosomes, or which carry out some special kind of metabolism; and from this point of view they would certainly seem to be much more than degenerate organs." Then I pointed out that not infrequently they are attached regularly to plasmosomes; and now I would call attention to the fact that they are still more frequently in contact with the nuclear membrane. Undoubtedly their function must be very different from that of the autosomes, because they appear and behave so different from them. The retention of the compact form and safraninophilous stain, so characteristic of many of them, throughout the growth period and in the rest stage of the spermatogonia, indicates that their nucleinic acid constituent changes less than in the autosomes. The sex determination by them, reasoned by McClung, Miss Stevens (1905b) and Wilson (1906), is a secondary function; if they do exercise a differentiation of sex this would be not their primal function but rather an indirect result of their metabolic peculiarities. From their position within the cell there can be little question that they fulfill an important part in the interplay of nuclear and cyto-plasmic activity, an influence perhaps in proportion to their size. Yet this influence can hardly be one of the nature of an assimilation process, else the chemical nature of the two allosomes could not remain so constant during the growth period.

2. The Nuclear Element and Chromosomal Difference.

More than twenty years ago Carnoy (1885) spoke of the Metazoan nucleus as con-taining an " élément nucléinien," by which he meant a continuous complex of linin and chromatin. We now know that his idea of nuclear structure was not exact, that, for instance, in the majority of nuclei there is no well marked chromatin spirem through the rest stage of the cell as he conceived it. Yet Carnoy had probably the right general idea. In my analysis of the spermatogenesis of Peripatus (1900), which was quite largely an examination of the changes of the linin threads, I went into con-siderable detail into the connection of the chromatin and the linin, and developed the thought very similar to that of Carnoy, that as the nuclear element of the first order should be considered the totality of the linin and chromatin. I conceived of this as a continuous and persisting linin band with which the chromatin masses are always in contact. The unity of this element is best seen in the prophases of cell division, where there is a continuous linin spirem with chromatin masses segregated upon it. But though the linin band becomes very much branched in the rest stage, and the chro-matin particles become finely distributed along these branches, yet there is consider-able evidence that it always maintains its continuity as a single band. In all sperma-togonic divisions the whole band, not only the chromatin masses, probably divides along its entire length, so that each daughter nucleus would receive one half of the original nuclear element; but in the reduction division this band would become transversely divided, therefore broken into as many portions as there are chromosomes. And I showed (1900, 1901b) that just after the reduction division, and in the earliest cleavages of the fertilized egg, the chromosomes are most distinct, presenting the appearance of small, independent vesicles. Therefore the reduction division causes the segmentation of the nuclear element, and accordingly it must become reconstituted before the spermatocyte and ovocyte stages of the next generation. All this is in accord with the phenomenon of the paternal and maternal chromosomes forming separate groups in the spindle in only the earlier embryonic cleavages, and not, as Häcker has argued, through the whole germinal cycle.

This was all elaborated at length in the earlier papers of mine referred to, and there shown to explain the mechanics of very diverse cellular changes. To that I would now add another thought. When the nuclear element becomes segmented by the reduction division, which is a division breaking the linin connections between conjugated chromosomes, its later reconstitution, *i. e.*, the restoration of a nuclear continuous nuclear element in the next generation, must take place by the maternal and paternal chromosomes arranging themselves in a continuous chain in such a way, that every two correspondent paternal and maternal chromosomes lie together. For this alone would explain why chromosomes of corresponding appearance are placed together in the prophases of division, and how in the synapsis stage of the growth period corresponding chromosomes conjugate unerringly.

The main results of these observations and interpretations amount to this, that the important nuclear element of the first order is a continuous band of linin with which chromatin is always locally connected. Beyond this there is in the nucleus nothing but the karyolymph, the nucleoli (plasmosomes), and minute floating granules (œdematin or lanthanin). With considerable justification we may assign to this nuclear element the main activities of heredity and differentiation, because it is the most constant structure.

Therefore we are to conceive of chromosomes not as separated nuclear masses, but as bodies in continuous physical connection. And each chromosome is a mass not of chromatin alone, but of chromatin always combined with linin, whether the chromatin be condensed as in mitosis, or whether it be finely distributed along delicate linin fibrils as in the rest stage. These two substances must be considered conjointly in any concept of the "hereditable substance," and not, as so many seem inclined to do, only the chromatin.

As elements of a second, lower grade we find the chromosomes. And we may define chromosome as a particular portion of the nuclear element on which the chromatin becomes massed during cell division. We can imagine the relation most simply in this way : there is a continuous linin band, on which chromatin is always suspended, more or less sparsely and irregularly when the cell is not in division, but in compact masses during division ; each portion of a linin band on which chromatin is so massed in division is a chromosome. Whether the movement of the chromatin particles on this band is automatic, or whether it is produced by local contractions of the linin, we have no means of deciding ; but certainly it is independent of extra-nuclear energies.

This idea of mine of the chromosomes as mere portions of a continuous nuclear element by no means implies that the chromosomes are not to be considered individuals, *i. e.*, structures that reappear in the same form and number in cell generation after generation. Indeed there is as much evidence that each chromosome is the product of a preceding one and not a new formation, as that a cell is always the division product of a preceding cell. And in all my work I have consistently argued for the chromosomes as persisting structures, in substantiation of the idea of the individuality of the chromosomes founded by Van Beneden, and supported by a great number of students.

Now in any consideration of the chromosomes the question presses on one : Are the several chromosomes of a given nucleus alike in their energies, or are they different? Are they actively or potentially equivalent, or are they not? Weismann and Roux were perhaps the first to take up this question, and Weismann has reasoned on the basis of his determinant hypothesis, that in any cell where the chromosomes are neither very small nor very numerous, each single chromosome is the bearer of all the hereditable qualities of a whole individual of the species. Against such a valence of the chromosome there is much evidence of serious weight, and it has been nowhere more succinctly summed up than in the recent review by Boveri (1904). To this matter of the potentiality of the chromosomes we will now turn.

Boveri has argued very strongly (1904) that particular chromosomes have particular energies, that one chromosome represents certain activities not evinced by another. His own important empirical contribution (1902) to this idea was the analysis of the abnormal development of eggs fertilized by one spermatozoön. And he concluded : "that not a fixed number but a fixed combination of chromosomes is necessary for normal development, and this means nothing else than that the particular chromosomes must possess different qualities."

Another line of evidence is that afforded by the differences in behavior of the chromosomes, when the cell is not molested by experiment. Such are the allosomes, of which we treated in the preceding section. They may behave differently from the autosomes, as we have seen, either by preserving their density in the rest period of the spermatogonia and the growth period of the spermatocytes, or by dividing in the maturation mitoses in a different sequence from the autosomes. Therefore in nuclei containing allosomes there are at least two kinds of chromosomes : the unmodified autosomes, and the modified allosomes; and there can be no doubt that these have different activities.

But we may go further than this. Are we to regard the possession of chromosomes of different kinds, particularly the possession of the highly modified allosomes, as simply a taxonomic peculiarity of certain forms, such as the insects, araneids, chilopods and *Sagitta?* I think not, for if there are such great differences in the chromosomes of these forms, is it not probable that there would be also chromosomal differences in other forms, even if less readily demonstrable?

For leaving the allosomes out of consideration comparative studies are proving dissimilarities of form and size in the unmodified chromosomes, the autosomes. I showed (1901b) that in a number of species of Hemiptera there are spermatogonic chromosome pairs marked by peculiarities in size; and that when this is the case there are corresponding bivalent elements in the first spermatocytes, i. e., that these size differences are constant during succeeding cell generations. I also showed in the same memoir that chromosomes of like size conjugate in the synapsis stage, and proved that of the two chromosomes that so conjugate the one is paternal and the other maternal, consequently that the synapsis is to be interpreted as the last stage in fertilization, the conjugation of correspondent chromosomes of opposite nativity. In the next year Sutton (1902) showed that in *Brachystola* all the autosomes compose pairs. And then (1904a) I demonstrated that in the spermatogonia of Urodelous Amphibia the twenty-four autosomes can be without difficulty resolved into twelve pairs, the components of a pair being distinguishable not only by size relations but also by peculiarities in form; and I showed this to be true of *Ascaris* also, where the ovotid contains one small and one large chromosome and the spermatozoön introduces one small and one large one. Wilson (1905) has recently found this to be the case for a number of Hemiptera, adding materially to my former observations; and in the present paper this constancy of pairs in the spermatogenesis is detailed for a still greater number of species. We can say that whenever the chromosomes are not too small or too numerous, they can be seen to present certain size relations that remain constant during succeeding cell generations, united sometimes with certain form relations as Baumgartner (1904) also has shown. McClung has likewise found this to hold true for certain of the Orthoptera.

So we are justified in saying that each spermatogonium and ovogonium has a double series of chromosomes, a paternal and a maternal set, which go to make up a series of pairs, the pairs being of gradated sizes or forms, and each pair composed of a paternal and maternal element of approximately equal size and form. The two elements of a pair probably lie close together in the spirem stage of the spermatogonium as I showed elsewhere (1904a); and even in the equatorial plate they frequently lie close together. The two elements of such a pair are the ones that conjugate in the synapsis stage, and that separate from each other in the first maturation division.

Accordingly, even where there are no such great differences present as between autosomes and allosomes, distinct pairs can frequently be distinguished, and thereby morphological differences of size and form be made out. It is obvious that chromosomes of different sizes cannot have the same physiological value; they must have activities differing at least in amount. But we may decide that their activities differ also in kind, else a particular chromosome would not always conjugate only with its correspondent in form and size but should be expected to conjugate with any other chromosome. That is to say, there is marked affinity or attraction only between the elements of such a pair, an attraction exhibited by the conjugation process. There is then something correspondent between the elements of a pair, not shared by them with the elements of any other pair, and this can be only a functional peculiarity, one based perhaps upon different metabolic energies. Therefore, as Sutton (1903) has reasoned, a chromosome must be the seat of particular qualities of the individual, not the center of the sum total of the individual's activities. Different chromosomes, that is to say, must have different physiological energies, and the sum of them, that is the whole nuclear element, present the energies of the individual.

Thus the experimental studies and the morphological ones are in accord in this matter, as Boveri (1904) has shown, and more recently Heider (1906). And these constant size and formal differences enable us to analyze the cell constituents much more fully than we could do a few years ago.

Another result I would mention here. When I first discovered the constancy of such chromosome pairs, I concluded that the two components of each pair were exactly equal in form and volume, and so have the others who followed me. In the present paper I have given especial attention to this point, and now find good evidence that the components of each pair are probably constantly slightly different from each other in volume. This is a difficult point to make sure of because it is hard to estimate voluminal mass in such small objects where there is much chance of optical illusion. But in most of those cases of pairs of small diplosomes of approximately equal volume, as those of the coreids, I find that they are always slightly different in volume in the first maturation mitosis; then always different in this respect in the spermatogonium; and here one can be fairly certain of his conclusion, because these bodies are nearly spherical and so relatively easy to compare. Again, in *Corizus alternatus* of the five pairs of autosomes of the spermatogonium, the largest pair (A, a, Fig. 107) is regularly composed of two relatively enormous elements, one slightly more voluminous and nearly straight, the other slightly smaller and horse-shoe shaped. And in *Harmostes*, where I have studied many spermatogonic divisions, all the autosome pairs are unusually distinct, and in each the two components appear constantly very slightly different in volume. This is clearly the case in *Ascaris* also. Now in this connection let us recall the discovery of Miss Stevens (1905b) and Wilson (1905a) that when there is a pair of diplosomes of markedly dissimilar volume, as in *Tenebrio* or *Euschistus*, the smaller must be the paternal element and the larger the maternal. If this is so for these diplosomes, is it not also probable that in any chromosome pair the slightly smaller element may be paternal and the larger one maternal? There would certainly seem to be a probability of this, and if it can be shown to be a constant relation it will give us the means of recognizing, after the determination of the chromosomal pairs, the maternal and paternal chromosomes of each nucleus, and thereby advance our means of analysis still another step.

And a word may be added here to those who may be sceptical as to the possibility of distinguishing particular chromosome pairs. Any one who looks over the plates given in this paper, and notes the chromosome pairs distinguished by corresponding letters, may say that the imagination plays too large a part in such distinctions. But he should recall that we can draw no conclusions without the help of the imagination, and that what we see we must also imagine. But more than this, he should recall that the printed figure can in no way be as clear as the preparation under the focussing microscope since it can reproduce only the profile, whereas the eye sees this and also the depth of the structure. One has only to draw the chromosomes carefully with the camera lucida, then search for correspondent ones upon such drawings, to be convinced of the actual presence of such pairs. And above all, no one has any right to express doubt of these relations who has not made broad comparative observations of his own.

This constant difference of the chromosome pairs, and the probable constant though much slighter differences of the elements of each pair, which are the expression of both morphological and physiological distinction, I would denote by the term "chromosome difference" which expresses the phenomena perhaps a little more precisely than Boveri's term "nuclear constitution."

3. The Number of Chromosomes and Taxonomy.

One incentive to me to make comparative studies of the chromosomes in the Hemiptera was to determine how far the number of chromosomes is constant in a particular group of animals; and certain conclusions were presented in two preceding papers (1901a, 1901b). From the observations on the Hemiptera then made it appeared that the chromosomal numbers were not constant, so that the determination of the factors governing the number seemed as unexplained as ever before. And in now touching on the question again I find that the problems are as difficult of solution as ever.

Yet it seemed worth while to reëxamine the matter from a taxonomic standpoint, to test the value of chromosome numbers as criteria of racial affinity. And since no one has tabulated the number of chromosomes known in animal species, not since the brief list of cases summarized by Wilson (1900; pp. 206, 207), I have compiled these statistics for the germ cells only of the greater number of described species; there are a number of omissions because some of the literature was inaccessible, but the list is very nearly complete. Data on hybrids are omitted; and data from certain older papers, as that of Carnoy (1885), where no particular pains were given to determining the numbers accurately, are left out. In the first vertical column of each table is given the name of the group, subgroup and species; in the second column the germinal cycle is indicated by the abbreviation "Ov" for ovogenesis, and "Sp" for spermatogenesis, in the third column are the names of the describers; and in the remaining columns the headings "Gonium," "Cyte I," "Cyte II," and "Tid" stand respectively for ovogonium (or spermatogonium), first ovocyte (or spermatocyte), second ovocyte (or spermatocyte), and ovotid (or spermatid). In these tables allosomes are not distinguished from autosomes since the intention is to present the entire chromosomal numbers. When a number is given as, *e. g.*, "10–11" it means that it was not determined whether 10 or 11 is present; but when it is stated "10, 11," it signifies that either 10 or 11 may be present, which of course would be a cycle complicated by the presence of a monosome. For the Hemiptera when my name is given as an authority, reference is made to the observations of the present paper.

Group and Species.	Cycle.	Authority.	Gonium	Cyte I	Cyte II	Tid.
VERTEBRATA.						
1. *Mammalia.*						
Bos taurus....................	Sp.	Schoenfeld, 1901.		12		
Lepus cuniculus..............	Ov.	Winiwarter, 1900.	ca. 42			
Mus rattus.....................	Sp.	Lenhossek, 1898.		12	12	12
Mus rattus.....................	"	Moore, 1894.	16	8	8	
Cavia cobaya..................	"	" 1906.	32	16	16	16
2. *Aves.*						
Columba livia.	Ov.	Harper, 1904.		8	8	8
3. *Amphibia.*						
Triton alpestris.........						
Triton cristatus.........	Sp.	Janssens, 1901.	24	12	12	12
Triton punctatus......						
Salamandra maculosa........	"	Meves, 1896; Janssens, 1901.	24	12	12	12
Batrachoseps attenuatus.....	"	Eisen, 1900; Janssens, 1903.	24	12	12	12
Desmognathus fusca.	"	Kingsbury, 1902; Montg.	24	12	12	12
Plethodon cinereus..........	"	Montg., 1904; Janssens, 1903.	24	12	12	12
Diemyctilus torosus..........	Ov.	Lebrun, 1901*b*.		12	12	12
Amphiuma means.	Sp.	McGregor, 1899.		12	12	12
Bufo lentiginosus.............	Ov.	King, 1901, 1905.	24	12	12	12
Rana temporaria	"	Lebrun, 1901*a*.		10	10	
4. *Pisces.*						
Myxine glutinosa	Sp.	Schreiner, 1905.	52	26	26	26
Salmo fario	Ov.	Böhm, 1892.		12	12	12
Scyllium canicula......						
Pristiurus	Sp.	Moore, 1895.	24	12	12	12
Torpedo						
Raja						

Group and Species.	Cycle.	Authority.	Gonium	Cyte I	Cyte II	Tid.
TUNICATA.						
Styelopsis grossularia..........	Ov.	Julin, 1893.	4	8	4	2
Styelopsis grossularia..........	Sp.	" "	4	4	2	1
Phallusia mammillata.........	Ov.	Hill, 1896.	8	8		? 8
Ascidia.	"	Boveri, 1890.	9			
ARACHNIDA.						
Agalena nævia	Sp.	L. B. Wallace, 1905.	40	20	19, 21	19, 21
Lycosa insopita	"	Montgomery, 1905.	26	13	13	
CHILOPODA.						
Scolopendra heros	"	Blackman, 1905.	33	17	16, 17	16, 17
Scutigera forceps	"	Medes, 1905.	37	19	18, 19	
INSECTA.						
1. Coleoptera.						
Dytiscus...........................	Ov.	Giardina, 1901.	ca. 40		10	10
Oryctes nasicornis	Sp.	Prowazek, 1901.	12	6	16	8
Tenebrio molitor	"	Stevens, 1905b.	20	10		
Hydrophilus	"	Vom Rath, 1892.	16	32	16	16
Cybister rœselii.......:........	"	Voinov, 1903.	ca. 22	13	12	12
Silpha carinata	"	Holmgren, 1902.	32	16	17	17
Agelastica alni :.................	Ov.	Henking, 1892.		12		
Agelastica alni	Sp.	" "	ca. 24	16–17	6–8	6–8
Donacia...........................	Ov.	" "		15	8	
Lampyris splendidula.........	"	" "		6–8		
Crioceris asparagi	"	" "				
2. Odonata.						
Anax junius	Sp.	McGill, 1904.	28	14	14	13, 14
3. Hymenoptera.						
Apis mellifica..................	Ov.	Petrunkewitsch, 1901.		16	16	8
Lasius niger	"	Henking, 1892.		10	10	
Rhodites rosæ	"	" "		ca. 9		
4. Isoptera.						
Termopsis angusticollis	Sp.	Stevens, 1905b.	52	26	26	26
5. Lepidoptera.						
Bombyx mori	"	Toyama, 1894.	26–28	26–28	28	14
Pieris brassicæ	Ov.	Henking, 1890a.		14	14	14
Pieris brassicæ	Sp.	" 1891.	30	14–15	14–15	14–15
Pieris napi......................	Ov.	" 1890b.	50	25		
6. Orthoptera.						
(Gryllidæ.)						
Gryllus assimilis...............	Sp.	Baumgartner, 1904.	29	15	14, 15	14, 15
Gryllus domesticus.............	"	" "	21	11	10, 11	10, 11
Gryllotalpa vulgaris...........	"	Vom Rath, 1892.	12	24	12	6
(Mantidæ.)						
Mantis religiosa	Sp.Ov.	Giardina, 1898.	14	14	14	7
(Blattidæ.)						
Periplaneta americana.........	Sp.	Moore and Robinson, 1905.	ca. 32	16	16	16
Blattella germanica.............	"	Stevens, 1905b.	23	12	11, 12	11, 12
(Locustidæ.)						
Orchesticus	"	McClung, 1902.	ca. 33	17	16, 17	16, 17
Orphania denticauda...........	"	de Sinéty, 1901.	31	16	15, 16	
Stenopelmatus...................	"	Stevens, 1905b.	46	24	23	24

Group and Species.	Cycle.	Authority.	Gonium	Cyte 1	Cyte II	Tid.
INSECTA (continued).						
7. *Hemiptera* (continued).						
(Phasmidæ.)						
Leptynia attenuata	"	de Sinéty, 1901.	36	19	18, 19	
(Forficulidæ.)						
Forficula auricularis	"	" "	24	12		
Labidura riparia	"	" . "		6		
(Acrididæ.)						
Brachystola magna.............	"	Sutton, 1902.	23	12	11, 12	11, 12
Caloptenus femur-rubrum....	"	Wilcox, 1895.	12	24	12	6
Hesperotettix speciosa.........	"	McClung, 1905.	23	11	11, 12	11, 12
Mermiria	"	" "	23	10	10	10, 11
Syrbula acuticornis	"	Montgomery, 1905.	20	10	10	10
7. *Hemiptera*.						
(Aphididæ.)						
Aphis rosæ.	Ov.Sp.	Stevens, 1905.	10	10	10	5
. Aphis œnotheræ.	"	" "	10	5	5	5
(Pentatomidæ.)						
Euschistus tristigmus	Sp.	Montgomery, Wilson, 1906.	14	8	7	7
Euschistus tristigmus	Ov.	Wilson, 1906.	14			
Euschistus variolarius.	Sp.	Montgomery, Wilson, 1906.	14	8	7	7
Euschistus variolarius.	Ov.	Wilson, 1906.	14	8		
Euschistus servus.	Sp.	" "	14		7	7
Euschistus servus.	Ov.	Wilson, 1906.	14		7	
Euschistus ictericus.	Sp.Ov.	" "	14		7	
Euschistus fissilis...............	Ov.	" "	14		8	
Euschistus fissilis.	Sp.	" 1905a.	14	8		
Mineus bioculatus.	"	" 1906.			7	7
Podisus spinosus...............	"	Montgomery, Wilson, 1905a.	16	9	8	8
Podisus spinosus...............	Ov.	Wilson, 1906.	16			
Mormidea lugens...............	Sp.	Montgomery.	14	8	7	
Cosmopepla carnifex.	"	" "	16	9	8	8
Nezara hilaris.	"	" Wilson, 1905a.	14	8	7	7
Nezara hilaris.	Ov.	Wilson, 1906.	14			
Brochymena......................	Sp.	Montgomery, Wilson, 1905a.	14	8	7	7
Perillus confluens...............	"	" "	14	8	7	7
Cœnus delius.....................	"	" Wilson, 1905a.	14	8	7	7
Cœnus delius.....................	Ov.	Wilson, 1906.	14			
Trichopepla semivittata	Sp.	" 1905a.	14	8	7	7
Trichopepla semivittata	"	Montgomery.	16	8	7	7
Eurygaster alternatus........ .	"	" "		7	6	6
Peribalus limbolaris............	"	" "	14	8	7	7
Banasa calva.....................	"	Wilson, 1905c.	15	13, 14	13, 14	
(Nabidæ.)						
Nabis annulatus.	"	Montgomery.		10	10	9
(Coreidæ.)						
Archimerus calcarator.	"	Wilson, 1905c.	15	8	7, 8	
Anasa tristis.....................	"	" 1905c, Montgomery.	21	11	11	10, 11
Anasa tristis.....................	Ov.	" 1906.	22			
Anasa armigera.	Sp.	Montgomery.	21	11	11.	
Anasa sp.	"	"	21	11		
Anasa sp.	Ov.	"	22			

Group and Species.	Cycle.	Authority.	Gonium	Cyte I	Cyte II	Tid.
INSECTA (continued).						
7. *Hemiptera* (continued).						
Harmostes reflexulus	Sp.	Montgomery, Wilson, 1906.	13		7	6, 7
Harmostes reflexulus	Ov.	Wilson, 1906.	14			
Corizus alternatus	Sp.	Montgomery.	13	7	7	6, 7
Corizus lateralis	"	"		7	7	6, 7
Chariesterus antennator	"	"		13	13	12, 13
Protenor belfragei	"	" Wilson, 1906.	13	7	7	6, 7
Protenor belfragei	Ov.	Wilson, 1906.	14			
Alydus pilosulus	Sp.	Montgomery, Wilson, 1905c.	13	7	7	6, 7
Alydus pilosulus	Ov.	Wilson, 1906.	14			
Alydus eurinus	Sp.	Montgomery.	13	7	7	6, 7
Metapodius terminalis	"	"	21	11	11	10, 11
Syromastes marginatus	"	Gross.	22	11	11	10, 11
(Lygæidæ.)						
Pyrrhocoris apterus	"	Henking, 1891.	24	12	12	11, 12
	Ov.	" 1892.	ca. 24	12	12	
Lygæus turcicus	Sp.	Wilson, 1905a.	14	8	7	7
Lygæus turcicus	Ov.	" 1906.	14			
Œdancala dorsalis	Sp.	Montgomery.	13	7	7	6, 7
Oncopeltus fasciatus	"	"	16	9	8	8
Peliopelta abbreviata	"	"	14	8, 9	7	7
Ichnodemus falicus	"	"	16	9	8	8
Cymus angustatus	"	"			14	? 14
(Tingitidæ.)						
Tingis clavata	Sp.	Montgomery.		7	7	7
(Phymatidæ.)						
Phymata	"	"	? 29			
(Reduviidæ.)						
Acholla multispinosa	"	"	32	16	16	
Sinea diadema	"	"		16		
Prionidus cristatus	"	"	26			
(Belostomatidæ.)						
Zaitha	"	"	24	13	12	12
(Hydrobatidæ.)						
Hygotrechus	"	"	21	11	11	10, 11
Limnotrechus marginatus	"	"		11	11	10, 11
(Capsidæ.)						
Calocoris rapidus	"	"	30	16	15, 16	
Pœcilocapsus goniphorus	"	"		18	17	17
Lygus pratensis	"	"	? 35	19	17, 18	17, 18
ONYCHOPHORA.						
Peripatus balfouri	"	" 1900.	ca. 28	14	14	14
CRUSTACEA.						
1. *Branchiopoda.*						
Artemia salina	Ov.	Brauer, 1893.	168	84	84	84, 168
Branchipus	Sp.	Moore, 1894.		10	10	ca. 5
Branchipus grubei	Ov.	Brauer, 1892.		12	12	12
2. *Copepoda.*						
Cyclops brevicornis	"	Häcker, 1902, 1904.	12	12	6	6
Cyclops strenuus	"	Rückert, 1894.		11	11	11

Group and Species.	Cycle.	Authority.	Gonium	Cyte I	Cyte II	Tid.
CRUSTACEA (continned).						
2. *Copepoda* (continued).						
Heterocope robusta	"	" "		16		
Diaptomus gracilis.............	"	" "		16		
Diaptomus	Ov.Sp.	Ishikawa, 1891.	8	8	8	4
3. *Isopoda.*						
Oniscus asellus..................	Sp.	Nichols, 1902.		16	16	16
4. *Ostracoda.*						
Cypris reptans.........	Ov.	Woltereck, 1890.		12		
5. *Decapoda.*						
Astacus............................	Sp.	Prowazek, 1902.		58		
BRACHIOPODA.						
Lingula anatina.................	Ov.	Yatsu, 1902.		8		
ENDOPROCTA.						
Pedicellina americana.........	Ov.Sp.	Dublin, 1905.	ca. 22	11	11	11
ECHINODERMATA.						
Strongylocentrotus.............	Ov.	Stevens, 1902.	36	18	18	18
Echinus esculentus............	"	Bryce, 1903.		16	16	16
Echinus microtuberculatus...	"	Boveri, 1905.	{ 18 { 36	9 18	9 18	9 18
Toxopneustes......................	"	Wilson, 1900.		18	18	18
PROSOPYGII.						
Phascalosoma	Ov.	Gerould, 1903.		10	10	
ANNELIDA.						
Thalassema mellita	"	Griffin, 1900.	24	12	12	12
Myzostoma glabrum............	"	Wheeler, 1897.		12	12	12
Allolobophora fœtida.........	"	Foot, 1898.	22	11	11	11
Ilyodrilus coccineus............	"	Vejdovský and Mrazek, 1903.		16		
Rhynchelmis....................	"	" "		32		
Ophryotrocha puerilis........	"	Korschelt, 1895.	4	4	2	2
Lumbricus	Sp.	Calkins, 1895.	32	16	16	16
Chætopterus pergamentaceus	Ov.	Mead, 1898.		9	9	9
Tomopteris.	"	W. Wallace, 1904.		4		
MOLLUSCA.						
1. *Gastropoda.*						
Enteroxenos östergreni.	"	Bonnevie, 1905.	34	17	17	17
(Prosobranchia.)						
Crepidula plana	"	Conklin, 1902.		30	30	30
Paludina vivipara.	Sp.	Meves, 1902.	14	7	7	7
Pterotrachea	Ov.	Boveri, 1890.		16	16	16
Carinaria.	"	" "		16	16	16
(Pulmonata.)						
Helix pomatia..................	Sp.	Ancel, 1903.	48	24	24	24
Helix pomatia.............	"	Prowazek, 1901b; v. Rath, 1896.	24	12	12	12
Helix pomatia..................	"	Lee, 1897.	24	24	24	24
Limax maximus.	Ov.	Linville, 1900.		?16		
Limax cinereo-niger.	Sp.	Vom Rath, 1892.	16	32	16	8
Limnæa elodes	Ov.	Linville, 1900.		16	16	

Group and Species.	Cycle.	Authority.	Gonium	Cyte I	Cyte II	Tid.
MOLLUSCA (continued).						
1. *Gastropoda* (continued).						
(Opisthobranchia.)						
Aplysia punctata	"	Janssens and Elrington, 1904.		16	16	16
Aplysia depilans	"	Bochenek, 1899.		16		
Haminea solitaria.	"	Smallwood, 1904.		16	16	
Phyllirhœ.	"	Boveri, 1890.		16	16	16
Cymbulia peronii	"	Nekrassoff, 1903.		16	16	16
2. *Pelecypoda.*						
Mactra.	"	Kostanecki, 1904.		16	16	16
CHÆTOGNATHA.						
Sagitta elegans ⎫ Sagitta bipunctata. ⎭	Ov.Sp.	Stevens, 1903, 1905c.	18	9	9	9
GORDIACEA.						
Paragordius varius	Ov.	Montgomery, 1904b.		7	7	7
Gordius.	"	Camerano, 1899.	ca. 8			
ACANTHOCEPHALA.						
Echinorhynchus gigas	Sp.	Kaiser, 1893.	4	4	4	2
NEMATODA.						
Ascaris megalocephala bivalens.	Ov.Sp.	Van Beneden, 1883; Hertwig, 1890; Boveri, 1887; Brauer, 1893.	4	2	2	2
Ascaris megalocephala univalens	"	Carnoy, 1886; Brauer, 1893; Boveri, 1887.	2	1	1	1
Ascaris sp. (from Canis)	Ov.	Lukjanow, 1889.		16	8	4
Ascaris sp. (from Canis)	Ov.	Carnoy, 1886.		8	4	
Ascaris lumbricoides.	"	Boveri, 1887.	24, 48			
Ascaris clavata	"	Carnoy, 1887.		24	24	24
Spiroptera strumosa	"	Carnoy, 1886.		8	4	2
Filaroides mustelarum	"	" "		8	4	4
Ophiostomum mucronatum .	"	" "		6	6	6
Strongylus tetracanthus	"	Meyer, 1895.		6		
NEMERTINI.						
Cerebratulus marginatus	"	Coe, 1899; Kostanecki, 1902.		16		
Tetrastemma vermiculum	"	Lebendinsky, 1897.		4	4	2
TURBELLARIA.						
1. *Polycladidea.*						
Prosthiostomum siphunculus	"	Francotte, 1898.		8	8	8
Leptoplana tremellaris	"	" 1897.		8	8	8
Oligocladus auritus	"	" "		8	8	8
Cycloporus papillosus	"	" "		8	8	8
Prostheceræus vittatus	"	Francotte, 1897; Gérard, 1901; Klinckowström, 1896.		6	6	6
Thysanozoön brocchi	"	Schockaert, 1902; Van der Stricht, 1898.	18	9	9	9
Eustylochus ellipticus	"	Van Name, 1899.		10	10	10
Planocera nebulosa	"	Van Name, 1899.		10	10	10
2. *Rhabdocœla.*						
Mesostomum ehrenbergi	"	Bresslau, 1904.		10	5	5

Group and Species.	Cycle.	Authority.	Gonium	Cyte I	Cyte II	Tid.
TURBELLARIA (continued).						
3. *Tricladidea.*						
Planaria simplicissima........	"	Stevens, 1904.			3	3
Planaria simplicissima........	Sp.	" "		8	4	4
"Freshwater forms".........	Ov.	Mattiesen, 1903.		8	4	4
TREMATODA.						
Polystomum integerrimum...	"	Goldschmidt, 1902.		8	8	4
Zoögonus mirus................	"	" 1905.	10	10	10	5
Gyrodactylus elegans.	"	Kathariner, 1904.		8	8	4
CNIDARIA.						
Hydra.............................	Sp.	Downing, 1905.	ca. 48	24	24	24
Æquorea forskalea.............	Ov.	Häcker, 1892.			6	6
Tiara..............................	"	Boveri, 1890.		14		
Gonothyrea lovenii	"	Wulfert, 1902.		8		
Clava squamata.................	"	Harm, 1902.	ca. 16			

For purposes of comparison the chromosomal numbers of the spermatogonia (and ovogonia), or those of the ovotids (and spermatids), are the safest to consider, because in cells of these generations in almost all cases the chromosomes are univalent, while different observers have varied greatly in their estimates of the valence of chromosomes of the ovocytes and spermatocytes. It is probable that the spermatogonic (or ovogonic) number of chromosomes is always double that of the number in the spermatid (or ovotid), so that the one can be readily calculated from the other; the only exception is in cases of spermatogenesis with a monosome, where the spermatid may contain one more chromosome or one less than half the number in the spermatogonium. And for purposes of comparison the full (not reduced) number of chromosomes is preferable, because in any species all the spermatogonia have the same number of chromosomes, while the spermatids may have different numbers.

Wilson's discovery that when there is an uneven spermatogonic number of chromosomes in the spermatogenesis there is an even number in the ovogenesis introduces a complexity in the comparisons. But this is easily obviated; for so far as known when the spermatogenesis has an uneven number it contains always one chromosome less than the ovogenesis, therefore, *e. g.*, a spermatogonium having 13 chromosomes we can calculate the ovogonium to have 14. In such cases we will use for comparison only the number of the ovogenesis, whether directly ascertained or whether derived by adding one to the spermatogonic number when the latter is an odd one.

When we look over the statistics presented in these tables we find that the number of chromosomes of the ovogonium or spermatogonium (translating odd spermatogonic numbers into even ovogonic ones) may be arranged in their order of frequency as follows :

24 chromosomes is the unreduced number in 30 species, about one-sixth of the whole list; the numbers 32 and 14 occur each in 24 species; the number 16 in 20 species ; the numbers 12 and 22 each in 9 species ; the numbers 18 and 20 each in 7 species ; the numbers 4, 8, 30 each in 6 species; the numbers 28 and 36 each in 5 species; the numbers 10, 34, 48 each in 4 species; the numbers 26, 40, 52 each in 2 species ; and the numbers 2, 38, 42, 46, 50, 60, 64, 116, 168, each in only one species.

Thus the full number of chromosomes is below 34 in the greater number of species so far studied.

Certain of these animals show the rare peculiarity of having two normal numbers, one twice that of the other ; thus *Ascaris megalocephala* has either 2 or 4, *Ascaris lumbricoides*, 24 or 48, *Helix pomatia*, 24 or 48, and *Echinus microtuberclatus*, either 18 or 36. In each of these species we might distinguish then a variety "univalens" from one "bivalens," as O. Hertwig (1890) has done for *Ascaris megalocephala*. In the last form Meyer (1895) was able to distinguish no anatomical differences between the two varieties, and Herla (1893) has proven that there is frequently crossing between them. But such hybrids contain three chromosomes, not twice the lower normal number. And evidently variation in the normal number, such as that of the four species mentioned, cannot have originated by polyspermy, for three spermatozoa would have to fertilize an ovum to produce double the usual normal number of chromosomes; and Boveri (1902) has shown that such polyspermy results in abnormal development.

Further, two cases are known where the spermatid has a different number of chromosomes from that of the ovotid, *Planaria* and *Styelopsis*, these being cases not due to the presence of a monosome in the spermatogenesis.

Finally, let us examine the constancy of the chromosomal numbers within certain circumscribed groups of animals. In some a certain constancy is to be found : the normal number is 24 in all the urodelous Amphibia ; McClung (1905) states there are always 23 in the spermatogenesis of the Acrididæ among the Orthoptera (but *Syrbula* and *Caloptenus* are exceptions to this) ; among the Pentatomidæ (Hemiptera) either 14 or 16 is the number (17 species examined), but *Banasa* has probably about 28 ; in the Coreidæ the numbers are 22 or 14 (one with 16) ; in all the opisthobranch molluscs examined it is 32 ; and in the Turbellaria, 12, 16 (most usual), 18 or 20. In most of the other groups of equivalent scope the variation in number is so great that there seems to be no constancy ; thus in the hemipteran family Lygæidæ there may be 24, 14, 16 or 28. And in the spermatogenesis of two closely related species of *Gryllus* Baumgartner (1904) finds the numbers to be 21 and 29.

We can decide this much about numerical relations of the chromosomes, that correspondence in number by no means implies community of race ; one has simply to list the different animals with the number 24 to be sure of this. On the other hand there is often constancy through smaller groups such as genera or species. The question is then : when we find a genus like *Ascaris*, with chromosomal number ranging from 2 to 48, are we to judge from this variation that chromosomal number has no taxonomic significance, or are we to decide that the forms combined in the genus *Ascaris* are really not generically related ?

This is an exceedingly difficult question to decide. If our present relegation of the species of *Ascaris* be justified, then clearly chromosomal numbers have not even generic worth. But our whole classification of somatic individuals is at present merely tentative, and the grouping of the various species of the Nematodes in particular seems to be very artificial. There is uncertainty at both ends of the argument. We must commence with the premise, that seems to me fully justified, that the species is one and the same from the egg up to the adult condition. Therefore it is permissible to classify germ cells as well as adults, and, *e. g.*, to compare chromosomal relations through a series of germ cells as we would conditions of the nervous system through a series of somatic individuals. The chromosomes as portions of the very conservative nuclear element should surely offer as good a basis for genetic comparisons as any set of somatic structures. That is to say, an entirely rational phylogeny of organisms might be founded in part upon relations of the germ cells ; therefore nuclear constituents be used as characters quite as much as any other sets of structures. The only reason to prefer comparisons of adult individuals is because they exhibit differentiation more than germ cells do, and not because they are really more differentiated.

Therefore when germ cells show differences in chromosomal numbers, these can signify only differences of the individuals that contain them. And while numerical differences are among the least important of the anatomical characters, yet when they are differences of so important an organ as the nuclear element they should be granted some degree of importance in a rational taxonomy. Consequently, it would be incorrect to place different species, some with 4 and others with 48 chromosomes, in the same genus, for such differences of the chromosomal number must constitute at least genetic and much more than specific difference. Were this not so, we could not explain why in so many cases there is constancy of chromosomal number in groups much higher than genera. Therefore chromosomal number is a character that should be considered in taxonomy.

At the same time number is only one of the properties of chromosomes, they have also other characters of form, arrangement, and process change, some of which will undoubtedly be found to be of greater value than number in the analysis of descent. McClung (1905) was the first to draw attention to arrangement of the chromosomes as a high taxonomic character, thus seconding my idea (1901a, b) that there should be a comparative phylogenetic study of the germ cells as a check and supplement to the analysis of the phylogeny based upon somatic structures. The foundation of a rational phylogenetic system upon cellular differences is as yet little more than suggested, because the comparative basis is so small and the phenomena so complex. Yet I believe it should be attempted, and that it will be found to be entirely possible.

Perhaps the best way of attacking the problem of the influences determining chromosomal number, is by the analysis of the phenomena in those species where there are two normal numbers.

In conclusion the position of the chromosomes in the equatorial plates of the maturation mitoses of the Hemiptera may be summarized.

Those diplosomes that divide equationally in the first mitosis and reductionally in the second are not central in the first spindle (except in *Oncopeltus*), but are central in the second spindle.

Those diplosomes that divide first reductionally and second equationally are always central in the first maturation spindle (except in the Reduviidæ), and more or less excentric in the second spindle (but central in the Reduviidæ).

It is therefore the rule that the positions of the diplosomes are reversed in the two maturation spindles; and that they are in the center of the chromosomal plate when they are bivalent (except in the Reduviidæ). Consequently the position of the diplosomes is rather a criterion of their valence than a character of any taxonomic importance.

There is a tendency in most of the Hemiptera, when the autosomes are not very numerous, for those of the first maturation spindle to be disposed in a circle around a central one, while there is generally less regularity in the second maturation spindle. Such positions would seem to be dependent upon the interaction of the number of chromosomes and the mechanics of the cell division, and therefore to be of no particular taxonomic importance.

LITERATURE LIST.

ANCEL, P., 1903. Réduction numérique des chromosomes dans la spermatogénèse d'Helix pomatia. Bibliogr. anat., 11.

BAUMGARTNER, W. J., 1904. Some new evidences for the individuality of the chromosomes. Biol. Bull., 8.

BLACKMAN, M. W., 1903. The spermatogenesis of the Myriapods. 2. On the chromatin in the spermatocytes of Scolopendra heros. Ibid., 5.

1905. Idem., 3. The spermatogenesis of Scolopendra heros. Bull. Mus. Comp. Zool. Harvard, 48.

BOCHENEK, A., 1889. Die Reifung und Befruchtung des Eies von Aplysia depilans. Bull. Acad. Cracovie.

BÖHM, 1892. Die Befruchtung des Forelleneies. Sitzber. Ges. Morph. Physiol. München, 7.

BONNEVIE, K., 1905. Das Verhalten des Chromatins in den Keimzellen von Enteroxenos östergreni. Anat. Anz., 26.

BOVERI, T., 1887. Zellen-Studien. 1. Die Bildung der Richtungskörper bei Ascaris megalocephala und Ascaris lumbricoides. Jena.

1890. Zellen-Studien. 3. Ueber das Verhalten der chromatischen Kernsubstanz bei der Bildung der Richtungs-körper und bei der Befruchtung. Jena. Zeitsch. Naturw., 24.

1902. Ueber mehrpolige Mitosen als Mittel zur Analyse des Zellkerns. Verh. phys.-med. Ges. Würzburg, N. F., 35.

1904. Ergebnisse über die Konstitution der chromatischen Substanz des Zellkerns. Jena.

1905. Zellen-Studien. 5. Ueber die Abhängigkeit der Kerngrösse und Zellenzahl der Seeigel-Larven von der Chromosomenzahl der Ausgangszellen. Jena.

BRAEM, F., 1897. Die geschlechtliche Entwickelung von Plumatella fungosa. Zoologica, Stuttgart, 23.

BRAUER, A., 1892. Das Ei von Branchipus Grubii von der Bildung bis zur Ablage. Abh. preuss. Akad. Wiss.

1893a. Zur Kenntniss der Reifung des sich parthenogenetisch entwickelnden Eies von Artemia salina. Arch. mikr. Anat., 43.

1893b. Zur Kenntniss der Spermatogenese von Ascaris megalocephala. Ibid., 42.

BRESSLAU, E., 1904. Beiträge zur Entwicklungsgeschichte der Turbellarien. 1. Die Entwickelung der Rhabdo-coelen und Alloiocolen. Zeit. wiss. Zool., 76.

BRYCE, T. H., 1903. Maturation of the ovum in Echinus esculentus. Quart. Journ. Micr. Sci. (N. S.), 46.

CALKINS, G. N., 1895. The spermatogenesis of Lumbricus. Journ. Morph., 11.

CARNOY, J. B., 1885. La cytodiérèse chez les Arthropodes. La Cellule, I.

1886. La cytodiérèse de l'oeuf. Ibid., 2, 3.

1887. Les globules polaires de l'Ascaris clavata. Ibid., 3.

COE, W. R., 1899. The maturation and fertilization of the egg of Cerebratulus. Zool. Jahrb., 12.

CONKLIN, E. G., 1902. Karyokinesis and cytokinesis in the maturation, fertilization and cleavage of Crepidula and other Gasteropoda. Journ. Acad. Nat. Sci. Philadelphia (2), 12.

DOWNING, E. R., 1905. The spermatogenesis of Hydra. Zool. Jahrb., 21.

DUBLIN, L. I., 1905. The history of the germ cells in Pedicellina americana. Ann. New York Acad. Sci., 16.

EBNER, V. v., 1900. Ueber die Theilung der Spermatocyten bei den Säugethieren. Sitzb. Akad. Wiss. Wien, 108.

EISEN, G., 1900. Spermatogenesis of Batrachoseps. Journ. Morph., 17.

FARMER, J. B., AND MOORE, J. E. S., 1905. On the maiotic pase (reduction divisions) in animals and plants. Quart. Journ. Micr. Sci. (N. S.), 48.

FOOT, K., 1898. The cocoons and eggs of Allolobophora foetida. Journ. Morph., 14.

FRANCOTTE, P., 1897. Recherches sur la maturation, la fécondation et la segmentation chez les Polyclades. Mém. couronnés Acad. roy. sci. Belg.

1898. Idem. — Arch zool. expér. génér.

GÉRARD, O., 1901. L'ovocyte de premier ordre du Prosthecerœus vittatus avec quelques observations relatives à la maturation chez trois autres Polyclades. La Cellule, 18.

GEROULD, J. H., 1903. The development of Phascolosoma. Arch. zool. expér. génér. (4), 2.

GIARDINA, A., 1898. Primi stadi embrionali della "Mantis religiosa." Monitore Zool. Italiano, 8.

1901. Origine dell' oocite e delle cellule nutrici nel Dytiscus. Internat. Monatsch. Anat. Phys., 18.

GOLDSCHMIDT, R., 1902. Untersuchungen über die Eireifung, Befruchtung und Zelltheilung bei Polystomum integerrimum Rud. Zeit. wiss. Zool., 71.

1905. Eireifung, Befruchtung und Embryonalentwicklung des Zoogonus mirus Lss. Zool. Jahrb., 21.

GRÉGOIRE, V., 1905. Les résultats acquis sur les cinèses de maturation dans les deux règnes. I. La Cellule, 22.

GRIFFIN, L. E., 1900. Studies on the maturation, fertilization, and cleavage of Thalassema and Zirphaea. Journ. Morph., 15.

GROSS, J., 1904. Die Spermatogenese von Syromastes marginatus L. Zool. Jahrb., 20.

GUYER, M. F., 1900. Spermatogenesis of normal and of hybrid pigeons. Chicago.

HÄCKER, V., 1892. Die Furchung des Eies von Aequorea Forskalia. Arch. mikr. Anat., 40.

1902. Ueber das Schicksal der elterlichen und grosselterlichen Kernanteile. Jena. Zeit. Naturwiss., 37.

1904. Bastardirung und Geschlechtszellenbildung. Zool. Jahrb. Supplement, 7.

HARM, K., 1902. Die Entwickelungsgeschichte von Clava squamata. Zeit. wiss. Zool., 73.

HARPER, E. H., 1904. The fertilization and early development of the pigeon's egg. Amer. Journ. Anat., 3.

HEIDER, K., 1906. Vererbung und Chromosomen. Jena.

HENKING, H., 1890a. Das Ei von Pieris brassicae L., nebst Bemerkungen über Samen und Samenbildung. Zeit. wiss. Zool., 49.

1890b. Ueber Reductionstheilung der Chromosomen in den Samenzellen von Insekten. Internat. Monatschr. Anat. Phys., 7.

1891. Ueber Spermatogenese und deren Beziehung zur Eientwicklung bei Pyrrhocoris apterus M. Zeit. wiss. Zool., 51.

1892. Untersuchungen über die ersten Entwicklungsvorgänge in den Eiern der Insekten. 3. Specielles und Allgemeines. Ibid., 54.

HERLA, V., 1893. Étude des variations de la mitose chez l'Ascaride mégalocéphale. Arch. de Biol., 13.

HERTWIG, O., 1890. Vergleich der Ei- und Samenbildung bei Nematoden. Arch. mikr. Anat., 36.

HILL, M. D., 1896. Notes on the fecundation of the egg of Sphaerechinus granularis, and on the maturation and fertilization of the egg of Phallusia mammillata. Quart. Journ. Micr. Sci. (N. S.), 38.

HOLMGREN, N., 1902. Ueber den Bau der Hoden und die Spermatogenese von Silpha carinata. Vorläufige Mittheilung. Anat. Anz., 22.

ISHIKAWA, 1891. Spermatogenesis, ovogenesis and fertilization in Diaptomus sp. Journ. Coll. Sci. Imper. Univ. Japan, 5.

JANSSENS, F. A., 1901. La spermatogénèse chez les Tritons. La Cellule, 19.

JANSSENS, F. A., AND DUMEZ, R., 1903. L'élement nucléinien pendant les cinèses de maturation des spermatocytes chez Batrachoseps attenuatus et Pletodon cinereus. Ibid., 20.

JANSSENS, F. A., AND ELRINGTON, G. A., 1904. L'élement nucléinien pendant les divisions de maturation dans l'oeuf de l'Aplysia punctata. Ibid., 21.

JULIN, C., 1893. Structure et développement des glandes sexuelles ; ovogénèse, spermatogénèse et fécondation chez Styelopsis grossularia. Bull. Sci. France et Belgique, 25.

KAISER, J. E., 1893. Die Acanthocephalen und ihre Entwickelung. Bibliotheca zoologica, 7.

KATHARINER, L., 1904. Ueber die Entwicklung von Gyrodactylus elegans v. Nrdm. Zool. Jahrb. Supplement, 7.

KING, H. D., 1901. The maturation and fertilization of the egg of Bufo lentiginosus. Journ. Morph., 17.

1905. The formation of the first polar spindle in the egg of Bufo lentiginosus. Biol. Bull., 9.

KINGSBURY, B. F., 1902. The spermatogenesis of Desmognathus fusca. Amer. Journ. Anat., 1.

KLINCKOWSTRÖM, A. v., 1896. Beiträge zur Kenntnis der Eireifung und Befruchtung bei Prosthecereaeus vittatus. Arch. mikr. Anat., 48.

KORSCHELT, E., 1895. Ueber Kerntheilung, Eireifung und Befruchtung bei Ophryotrocha puerilis. Zeit. wiss. Zool., 60.

KOSTANECKI, A., 1902. Ueber de Reifung und Befruchtung des Eies von Cerebratulus marginatus. Bull. Acad. Sci. Cracovie.

1904. Cytologische Studien an künstlich parthenogenetisch sich entwickelnden Eiern von Mactra. Arch. mikr. Anat., 64.

LEBENDINSKY, J., 1897. Zur Entwickelungsgeschichte der Nemertinen. Biol. Centralbl., 17.

LEBRUN, H., 1901a. La vésicule germinative et les globules polaires chez les Anoures. La Cellule, 19.

1901b. Les cinéses sexuelles chez Diemyctilus torosus. Ibid., 20.

LÉCAILLON, A., 1901. Recherches sur l'ovaire des Collemboles. Arch. Anat. Micr. Paris, 4.

LEE, A. B., 1897. Les cinéses spermatogénetiques. La Cellule, 13.

LENHOSSÉK, M. v., 1898. Untersuchungen über Spermatogenese. Arch. mikr. Anat., 51.

LINVILLE, H. R., 1900. Maturation and fertilization in Pulmonate Gasteropods. Bull. Mus. Com. Zool. Harvard.

LUKJANOW, S. M., 1889. Einige Bemerkungen über sexuelle Elemente beim Spulwurm des Hundes. Arch. mikr. Anat., 34.

McCLUNG, C. E., 1902. The spermatocyte divisions of the Locustidæ. Kansas Univ. Sci. Bull., 1.

1905. The Chromosome complex of Orthopteran spermatocytes Biol. Bull., 9.

McGILL, C., 1904. The Spermatogenesis of Anax junius. Univ. Missouri Studies, 2.

McGREGOR, J. H., 1899. The Spermatogenesis of Amphiuma. Journ. Morph.

MATTIESEN, E., 1903. Die Eireifung und Befruchtung der Süsswasserdendrocoelen. Zool. Anz., 27.

MEAD, A. D., 1898. The origin and behavior of the centrosomes in the Annelid egg. Journ. Morph., 14.

MEDES, G., 1905. The Spermatogenesis of Scutigera forceps. Biol. Bull., 9.

MEVES, F., 1896. Ueber die Entwickelung der männlichen Geschlechtszellen von Salamandra maculosa. Arch. mikr. Anat., 48.

1902. Ueber oligopyrene und apyrene Spermien und über ihre Entstehung, nach Beobachtungen an Paludina und Pygæra. Ibid. 61.

MEYER, O., 1895. Celluläre Untersuchungen an Nematoden-Eiern. Jena. Zeitsch. Naturw., 29.

MOENKHAUS, W. J., 1904. The development of the hybrids between Fundulus heteroclitus and Menidia notata. Amer. Jour. Anat., 3.

MONTGOMERY, T. H., JR., 1898. The spermatogenesis in Pentatoma up to the formation of the spermatid. Zool. Jahrb., 12

1900. The spermatogenesis of Peripatus (Peripatopsis) balfouri up to the formation of the spermatid. Ibid., 14.

1901a. Further studies on the chromosomes of the Hemiptera heteroptera. Proc. Acad. Nat. Sci. Philadelphia.

1901b. A study of the chromosomes of the germ cells of Metazoa. Trans. Amer. Phil. Soc., 20.

1904a. Some observations and considerations upon the maturation phenomena of the germ cells. Biol. Bull., 6.

1904b. The development and structure of the larva of Paragordius. Proc. Acad. Nat. Sci. Philadelphia.

1905. The spermatogenesis of Syrbula and Lycosa, with general considerations upon chromosome reduction and the heterochromosomes. Ibid.

1906. The terminology of aberrant chromosomes and their behavior in certain Hemiptera. Science (N. S.), 23.

MOORE, J. E. S., 1894. Some points in the spermatogenesis of Mammals. Internat. Monatschr. Anat. Phys., 11.

1894b. Some points in the origin of the reproductive elements in Apus and Branchipus. Quart. Journ. Micr. Sci. (N. S.), 35.

1895. On the structural changes in the reproductive cells during the spermatogenesis of Elasmobranchs. Ibid., 38.

MOORE, J. E. S., AND ROBINSON, L. E., 1905. On the behavior of the nucleolus in the spermatogenesis of Periplaneta americana. Ibid., 48.

MOORE, J. E. S. AND WALKER, C. E., 1906. The maiotic process in Mammalia. University Press of Liverpool.

NAME, W. G. VAN, 1899. The maturation, fertilization and early development of the Planarians. Trans. Conn. Acad. Sci., 10.

NEKRASSOFF, A., 1903. Untersuchungen über die Reifung und Befruchtung des Eies von Cymbulia peronii. Anat. Anz., 24.

NICHOLS, M. L., 1902. The spermatogenesis of Oniscus asellus Linn. Proc. Amer. Phil. Soc., 41.

PETRUNKEWITSCH, A., 1901. Die Richtungskörper und ihr Shicksal im befruchteten und unbefruchteten Bienenei. Zool. Jahrb., 14.

PROWAZEK, S., 1901a. Spermatogenese des Nashornkäfers (Oryctes nasicornis L.). Arbeit. zool. Inst. Wien., 13.

1901b. Spermatogenese der Weinbergschnecke (Helix pomatia L.). Ibid.

1902. Ein Beitrag zur Krebsspermatogenese. Zeit. wiss. Zool., 71.

RATH, O. VOM, 1892. Zur Kenntniss der Spermatogenese von Gryllotalpa vulgaris. Arch. mikr. Anat., 40.

1896. Neue Beiträge zur Frage der Chromatinreduction in der Samen- und Eireife. Ibid., 46.

RÜCKERT, J., 1894. Zur Eireifung bei Copepoden. Anat. Hefte.

SCHOCKAERT, R., 1902. L'ovogénèse chez le Thysanozoon brocchi. La Cellule, 20.

SCHOENFELD, H., 1901. La spermatogénèse chez le taureau et chez les mammifères en général. Arch. de Biol., 18.

SCHREINER, A., AND K. E., 1905. Ueber die Entwickelung der männlichen Geschlechtszellen von Myxine glutinosa (L.). Ibid., 21.

SINÉTY, R. DE, 1901. Recherches sur la biologie et l'anatomie des Phasmes. La Cellule.

SMALLWOOD, W. M., 1904. The maturation, fertilization and early cleavage of Haminea solitaria (Say). Bull. Mus. Comp. Zool. Harvard, 45.

STEVENS, N. M., 1902. Experimental studies on eggs of Echinus microtuberculatus. Arch. Entwicklungsmech., 15.

1903. On the ovogenesis and spermatogenesis of Sagitta bipunctata. Zool. Jahrb., 18.

1904. On the germ cells and the embryology of Planaria simplicissima. Proc. Acad. Nat. Sci. Philadelphia.

1905a. A study of the germ cells of Aphis rosæ and Aphis œnotheræ. Journ. Exper. Zool., 2.

1905b. Studies in spermatogenesis with especial reference to the "accessory chromosome." Carnegie Institute Publ.

1905c. Further studies on the ovogenesis of Sagitta. Zool. Jahrb., 21.

STRICHT, O. VAN DER, 1898. La formation des deux globules polaires et l' apparition des spermocentres dans l'oeuf de Thysanozoon Brocchi. Arch. de Biol., 15.

SUTTON, W. S., 1902. On the morphology of the chromosome group in Brachystola magna. Biol. Bull., 4.

1903. The chromosomes in heredity. Ibid.

TOYAMA, K., 1894. On the spermatogenesis of the Silk Worm. Bull. Coll. Agric. Imper. Univ. Japan, 2.

VEJDOVSKÝ, F., UND MRAZEK, A., 1903. Umbildung des Cytoplasma während der Befruchtung und Zellteilung. Arch. mikr. Anat., 62.

VOINOV, D., 1903. La spermatogénèse d'été chez le Cybister Roeselii. Arch. zool. expér. génér. (4), 1.

WALDEYER, W., 1888. Ueber Karyokinese und ihre Beziehungen zu den Befruchtungsvorgängen. Arch. mikr. Anat., 32.

WALLACE, L. B., 1905. The spermatogenesis of the Spider. Biol. Bull., 8.

WALLACE, W., 1904. The oocyte of Tomopteris. Report 73 Meet. Brit. Assoc. Adv. Sci.

WHEELER. W. M., 1897. The maturation, fecundation and early cleavage of Myzostoma glabrum Leuckart. Arch. de Biol., 15.

WILCOX, E. V., 1895. Spermatogenesis of Caloptenus femur-rubrum and Cicada tibicen. Bull. Mus. Comp. Zool. Harvard, 27.

WILSON, E. B., 1900. The cell in development and inheritance. 2d ed. New York.

1905a. Studies on Chromosomes. 1. The Behavior of the idiochromosomes in Hemiptera. Journ. Exper. Zool., 2.

1905b. The chromosomes in relation to the determination of sex in Insects. Science (N. S.), 22.

1905c. Studies on chromosomes. 2. The paired microchromosomes, idiochromosomes and heterotropic chromosomes in Hemiptera. Journ. Exper. Zool., 2.

1906. Idem. 3. The sexual differences of the chromosome-groups in Hemiptera, with some considerations of the determination and inheritance of sex. Ibid., 3.

WINIWARTER, H. V., 1900. Recherches sur l'ovogénèse et l'organogénèse de l'ovaire des Mamiféres (Lapin et Homme). Arch. de Biol., 17.

WOLTERECK, R., 1898. Zur Bildung und Entwickelung des Ostracoden-Eies. Zeit. Wiss. Zool., 64.

WULFERT, J., 1902. Die Embryonalentwickelung von Gonothyrea lovenii Allm. Ibid., 71.

Yatsu, N., 1902. On the development of Lingula anatina. Journ. Coll. Sci. Japan, 17.

UNIVERSITY OF TEXAS, March 26, 1906.

EXPLANATION OF THE PLATES.

All the figures have been drawn by the author with the camera lucida at the level of the base of the microscope, and the reproductions are the size of the originals. Figs. 1–68, 94–106 and 126–133 are drawn at a magnification of about 2,080 diameters, all the others at a magnification of about 2,480 diameters. Lateral views of the first maturation spindles are placed the length of the plate, of the second maturation spindle the width of the plate, which enables one to distinguish them at a glance. The following abbreviations have been employed:

Di, diplosome.

Mo, monosome.

Pl, plasmosome (true nucleolus).

The diplosomes are paired elements, and when their separate components can be distinguished, they are lettered *Di* and *di* respectively; in case there is more than one pair of them to a cell a number is placed after letters, viz., *Di. 1, di. 1* would be one pair and *Di. 2, di. 2* a second pair; the capital letter is used for the small component of a pair and the small letter for the larger one in those cases where they differ in size. If there is a single monosome present it is lettered simply *Mo*, but if two they are lettered *Mo. 1* and *Mo. 2*. Single letters denote autosomes, a capital and a lower case letter of the same kind (as *A* and *a*) marking the components of a pair; if the capital and the small letter are separated by a comma, as "*A, a*," a pair of correspondent ones is denoted; but if a capital is followed by a small letter enclosed in parentheses, as "*A (a)*," it is indicated that but one of the elements is present, *i. e.*, either *A* or *a*.

Some of the figures are redrawings of cells previously figured by me, and in such cases this is noted by the date of the paper where the particular cell was first illustrated followed by the number of the original figure, all this being enclosed in parentheses, as "(v. 1901*b*, Fig. 2)."

PLATE IX.

Figs. 1–14, *Euschistus variolarius*.

Figs. 1–4, spermatogonic monasters (with Fig. 1, v. 1901*b*, Fig. 2).

Fig. 5, nucleus in synapsis stage.

Figs. 6–9, successive prophases of the maturation mitosis, the last two showing all the chromosomes.

Fig. 10, first maturation monaster.

Figs. 11, 12, second maturation monasters.

Figs. 13, 14, chromosomes of two spermatids.

Figs. 15–22, *Euschistus tristigmus*.

Fig. 15, spermatogonic monaster.

Fig. 16, nucleus of synapsis stage.

Fig. 17, pole view of first maturation spindle.

Fig. 18, lateral view of the same.

Fig. 19, pole view of a plate of daughter elements before their arrangement in the spindle.

Fig. 20, second maturation spindle.

Figs. 21, 22, chromosomes of two spermatids.

Figs. 23–27, *Podisus spinosus*.

Fig. 23, spermatogonic monaster (v. 1901*b*, Fig. 27).

Fig. 24, nucleus of late synapsis stage.

Fig. 25, oblique lateral view of first maturation spindle.

Fig. 26, pole view, second maturation spindle.

Fig. 27, lateral view of the same stage.

Figs. 28-30, *Mormidea lugens.*

Fig. 28, spermatogonic monaster (v. 1901*b*, Fig. 31) ; the autosomes *C* and *c* are seen from their ends, and it could not be decided whether *E*, *e* is one or two elements.

Fig. 29, pole view of first maturation spindle.

Fig. 30, pole view of second maturation spindle.

Figs. 31-37, *Cosmopepla carnifex.*

Fig. 31, spermatogonic monaster.

Fig. 32, late prophase of first maturation division.

Fig. 33, pole view, first maturation spindle (v. 1901*b*, Fig. 41).

Fig. 34, lateral view, first maturation spindle (v 1901*b*, Fig. 40).

Figs. 35, 36, pole views, second maturation spindle.

Fig. 37, chromosomes of a spermatid.

Figs. 38, 39, *Nezara hilaris.*

Fig. 38, spermatogonic monaster (v. 1901*b*, Fig. 44).

Fig. 39, nucleus of postsynapsis stage.

Figs. 40-45, *Brochymena* sp.

Figs. 40, 41, spermatogonic monasters (with 41 v. 1901*b*, Fig. 47).

Fig. 42, oblique lateral view, first maturation spindle.

Fig. 43, pole view, first maturation spindle.

Fig. 44, second maturation spindle.

Fig. 45, pole view of the same stage.

Figs. 46-52, *Perillus confluens.*

Fig. 46, spermatogonic monaster.

Fig. 47, plasmosome and diplosomes of the early prophase of the first maturation division.

Fig. 48, pole view, first maturation spindle.

Fig. 49, pole view of a daughter plate of the first maturation mitosis.

Fig. 50, pole view, second maturation spindle.

Fig. 51, lateral view of the same stage (one of the elements not shown).

Fig. 52, chromosomes of a spermatid.

Figs. 53-58, *Cœnus delius.*

Figs. 53, 54, spermatogonic monasters.

Fig. 55, daughter plate of spermatogonic division.

Fig. 56, daughter plate, first maturation division.

Fig. 57, pole view, second maturation spindle.

Fig. 58, lateral view of the same stage.

Figs. 59-65, *Trichopepla semivittata.*

Fig. 59, spermatogonic monaster (v. 1901*b*, Fig. 65).

Figs. 60, 61, spermatocytic nuclei, late growth period.

Fig. 62, spermotocytic nucleus, rest stage.

Fig. 63, idem, early prophase of first maturation division.

Figs. 64, 65, first maturation spindles.

PLATE X.

Figs. 66-68, *Trichopepla semivittata.*

Fig. 66, second maturation spindle.

Fig. 67, pole view of the same stage.

Fig. 68, chromosomes of a spermatid.

Figs. 69–73, *Eurygaster alternatus*.

Fig. 69, spermatocyte nucleus, late postsynapsis.

Fig. 70, first maturation spindle, the chromosomes not in definite arrangement.

Fig. 71, pole view of the same stage.

Fig. 72, pole view, second maturation spindle.

Fig. 73, idem, lateral view.

Figs. 74–80, *Peribalus limbolaris*.

Fig. 74, spermatogonic monaster.

Fig. 75, spermatocyte nucleus, near end of growth period.

Fig. 76, pole view, first maturation spindle (v. 1901b, Fig. 37).

Fig. 77, oblique lateral view of the same stage.

Fig. 78, pole view, second maturation spindle.

Fig. 79, daughter plate, first maturation division.

Fig. 80, oblique lateral view, second maturation spindle.

Figs. 81–93, *Nabis annulatus*.

Figs. 81–85, successive prophases, first maturation division.

Fig. 86, pole view, first maturation spindle (v. 1901a, Fig. 14).

Figs. 87, 88, lateral views, first maturation spindle.

Fig. 89, pole view of early daughter plate of preceding division.

Figs. 90, 91, second maturation spindles.

Figs. 92, 93, chromosome plates of spermatids.

Figs. 94–106, *Harmostes reflexulus*.

Figs. 94, 95, spermatogonic monasters.

Fig. 96, spermatocyte nucleus, synapsis.

Fig. 97, idem, postsynapsis.

Fig. 98, idem, later postsynapsis.

Fig. 99, idem, rest stage.

Figs. 100, 101, idem, early prophases of first maturation division.

Figs. 102, 103, first maturation spindles (v. 1901 b, Figs. 113, 116).

Fig. 104, second maturation spindle.

Figs. 105, 106, chromosome plates of spermatids.

Figs. 107–116, *Corizus alternatus*.

Fig. 107, spermatogonic monaster (v. 1091a, Fig. 18).

Fig. 108, spermatocyte nucleus, late synapsis.

Figs. 109, 110, idem, postsynapsis.

Fig. 111, idem, rest stage.

Figs. 112, 113, the autosome A, a, prophase of first maturation division.

Figs. 114–116, successive prophases, first maturation division.

PLATE XI.

Figs. 117–122, *Corizus alternatus*.

Fig. 117, pole view, first maturation spindle.

Fig. 118, lateral view of the same stage.

Fig. 119, daughter chromosomal plate of preceding stage.

Fig. 120, pole view, second maturation spindle.

Figs. 121, 122, second maturation spindles.

A. P. S.—XXI. S. 27, 8, '06.

Figs. 123-125, *Corizus lateralis.*

Fig. 123, first maturation spindle.

Fig. 124, second maturation spindle.

Fig. 125, late anaphase of second maturation.

Figs. 126-133, *Chariesterus antennator.*

Fig. 126, spermatocyte nucleus, postsynapsis.

Figs. 127, 128, pole views, first maturation spindle.

Fig. 129, lateral view of the same stage.

Fig. 130, pole view, second maturation spindle.

Figs. 131, 132, corresponding daughter plates of second maturation division.

Fig. 133, anaphase, second maturation division.

Figs. 134, 135, *Protenor belfragei.*

Fig. 134, spermatogonic monaster (v. 1901b, Fig. 119).

Fig. 135, spermatocyte nucleus, late growth period.

Figs. 136-143, *Alydus pilosulus.*

Fig. 136, spermatogonic monaster.

Fig. 137, spermatocyte nucleus, early synapsis.

Fig. 138, idem, late synapsis.

Fig. 139, late prophase of first maturation division.

Fig. 140, first maturation spindle.

Figs. 141-143, successive second maturation spindles.

Figs. 144-150, *Alydus eurinus.*

Fig. 144, spermatogonic monaster (v. 1901b, Fig. 96).

Fig. 145, spermatocyte nucleus, late synapsis.

Fig. 146, pole view, first maturation spindle.

Fig. 147, lateral view of first maturation spindle.

Fig. 148, daughter plate, early anaphase, first maturation division.

Fig. 149, pole view, second maturation spindle.

Fig. 150, lateral view, second maturation spindle.

Figs. 151-161, *Anasa tristis.*

Fig. 151, spermatogonic monaster.

Figs. 152, 153, spermatocyte nuclei, synapsis stage.

Figs. 154, 155, idem, later growth period.

Fig. 156, pole view, first maturation spindle.

Figs. 157, 158, first maturation spindles.

Fig. 159, pole view, second maturation spindle.

Figs. 160, 161, second maturation spindles.

Figs. 162-166, *Anasa sp.*

Figs. 162, 163, ovogonic monasters.

Fig. 164, spermatogonic monaster.

Figs. 165, 166, pole and lateral views, first maturation spindle.

Figs. 167, 168, *Anasa armigera.*

Fig. 167, spermatogonic monaster.

Fig. 168, first maturation spindle.

Figs. 169, 170, *Metapodius terminalis.*

Figs. 169, 170, spermatogonic monasters (with 169 v. 1901d, Fig. 85).

PLATE XII.

Figs. 171–182, *Metapodius terminalis.*

Fig. 171, spermatocyte nucleus, synapsis.

Fig. 172, idem, postsynapsis.

Fig. 173, idem, rest stage.

Fig. 174, late prophase of first maturation division.

Fig. 175, first maturation spindle.

Figs. 176, 177, pole views of the same spindle.

Fig. 178, anaphase of the first maturation division.

Fig. 179, daughter plate, early anaphase, first maturation division.

Fig. 180, pole view, second maturation division.

Figs. 181, 182, second maturation spindles.

Figs. 183–195, *Œdancala dorsalis.*

Fig. 183, spermatogonic monaster (v. 1901*b*, Fig. 154).

Fig. 184, spermatocyte nucleus just before rest period.

Fig. 185, idem, rest stage.

Figs. 186–188, successive prophases of first maturation division.

Fig. 189, pole view, first maturation spindle.

Figs. 190, 191, first maturation spindles (with 191 v. Fig. 158, 1901*b*).

Fig. 192, pole view, second maturation spindle.

Fig. 193, second maturation spindle (v. 1901*b*, Fig. 157).

Fig. 194, second maturation anaphase.

Fig. 195, pole view of chromosomes of a spermatid.

Figs. 196–199, *Oncopeltus fasciatus.*

Fig. 196, daughter plate, early anaphase of first maturation division (v. 1901*b*, Fig. 171).

Figs. 197, 198, pole and lateral view, second maturation spindle.

Fig. 199, chromosome plate of a spermatid.

Figs. 200–210, *Peliopelta abbreviata.*

Fig. 200, spermatogonic monaster.

Fig. 201, idem (v. 1901*b*, Fig. 149).

Fig. 202, spermatocyte nucleus, synapsis.

Fig. 203, idem, postsynapsis.

Figs. 204, 205, late prophases, first maturation division.

Fig. 206, pole view, first maturation spindle.

Fig. 207, oblique lateral view of chromosomes of the same division.

Fig. 208, first maturation spindle.

Fig. 209, pole view, second maturation spindle.

Fig. 210, second maturation spindle.

Figs. 211–225, *Ichnodemus falicus.*

Fig. 211, spermatogonic monaster (v. 1901*b*, Fig. 145).

Figs. 212, 213, spermatocyte nuclei, postsynapsis.

Fig. 214, idem, end of growth period.

Figs. 215–219, successive prophases, first maturation division.

Fig. 220, first maturation spindle.

Figs. 221, 222, pole views of first maturation spindles (v. 1901*b*, Figs. 147, 148).

Fig. 223, second maturation spindle.

Fig. 224, pole view, second maturation spindle.

Fig. 225, chromosomes of a spermatid.

Figs. 226–228, *Cymus angustatus.*

Fig 226, pole view, second maturation spindle (v. 1901b, Fig. 144).

Figs. 227, 228, second maturation spindles.

PLATE XIII.

Figs. 229–236, *Tingis clavata.*

Fig. 229, pole view, first maturation spindle.

Fig. 230, oblique lateral view of the same stage.

Fig. 231, pole view, first maturation spindle.

Figs. 232–234, pole views, second maturation spindles.

Figs. 235, 236, chromosome plates of spermatids.

Fig. 237, *Phymata sp.*

Fig. 237, spermatogonic monaster (v. 1901b, Fig. 200).

Figs. 238–244, *Acholla multispinosa.*

Fig. 238, spermatogonic monaster (v. 1901b, Fig. 207).

Fig. 239, spermatocyte nucleus, rest stage.

Fig. 240, first maturation spindle.

Figs. 241, 242, pole views, first maturation spindle.

Figs. 243, 244, pole views, second maturation spindles (with 243 v. 1901b, Fig. 211).

Figs. 245–250, *Sinea diadema.*

Fig. 245, spermatocyte nucleus, rest stage.

Fig. 246, pole view, first maturation spindle.

Fig. 247, oblique lateral view of chromosomes, first maturation spindle.

Figs. 248–250, first maturation spindles (v. 1901b, Figs. 217, 218).

Figs. 251, 252, *Prionidus cristatus.*

Fig. 251, spermatogonic monaster (v. 1901b, Fig. 224).

Fig. 252, spermatocyte nucleus, rest stage.

Figs. 253–258, *Zaitha sp.*

Fig. 253, spermatogonic monaster.

Fig. 254, spermatocyte, rest stage.

Fig. 255, first maturation spindle.

Fig. 256, pole view, first maturation spindle.

Figs. 257, 258, lateral and pole views, second maturation spindle.

Figs. 259–268, *Hygotrechus sp.*

Fig. 259, spermatogonic monaster (v. 1901b, Fig. 229).

Fig. 260, spermatocyte nucleus, rest stage.

Fig. 261, late prophase, first maturation division.

Figs. 262–264, lateral and pole views, first maturation spindle (with 264 v. 1901b, Fig. 231).

Figs. 265–267, lateral and pole views, second maturation spindle.

Fig. 268, chromosome plate of a spermatid.

Figs. 269–274, *Limnotrechus marginatus.*

Fig. 269, spermatocyte, nucleus, rest stage.

Fig. 270, monosome and plasmosome, early prophase of first maturation division.

Figs. 271, 272, pole and lateral view, first maturation spindle (v. 1901b, Fig. 233).

Figs. 273, 274, lateral and pole view, second maturation spindle.

Figs. 275-287, *Calocoris rapidus.*

Fig. 275, spermatogonic monaster (v. 1901*b*, Fig. 177).

Fig. 276, spermatocyte nucleus, synapsis.

Fig. 277, idem, end of growth period.

Fig. 278, late prophase, first maturation division.

Figs. 279, 280, pole views, first maturation spindle (with 279 v. 1901*b*, Fig. 185).

Figs. 281-283, first maturation spindles (with 281 v. 1901*b*, Fig. 182).

Fig. 284, second maturation spindle.

Figs. 285, 286, pole views of second maturation spindles.

Fig. 287, second maturation anaphase.

Figs. 288-294, *Pœcilocapsus goniphorus.*

Fig. 288, spermatocyte nucleus, rest stage.

Fig. 289, pole view, first maturation spindle (v. 1901*b*, Fig. 196).

Fig. 290, first maturation spindle.

Fig. 291, pole view, second maturation spindle (v. 1901*b*, Fig. 197).

Fig. 292, second maturation spindle.

Figs. 293, 294, corresponding daughter plates, early anaphase of second maturation division.

Figs. 295-299, *Lygus pratensis.*

Figs. 295, 296, pole and lateral views, first maturation spindle.

Figs. 297, 298, pole and lateral views, second maturation division.

Figs. 299, anaphase of second maturation division.

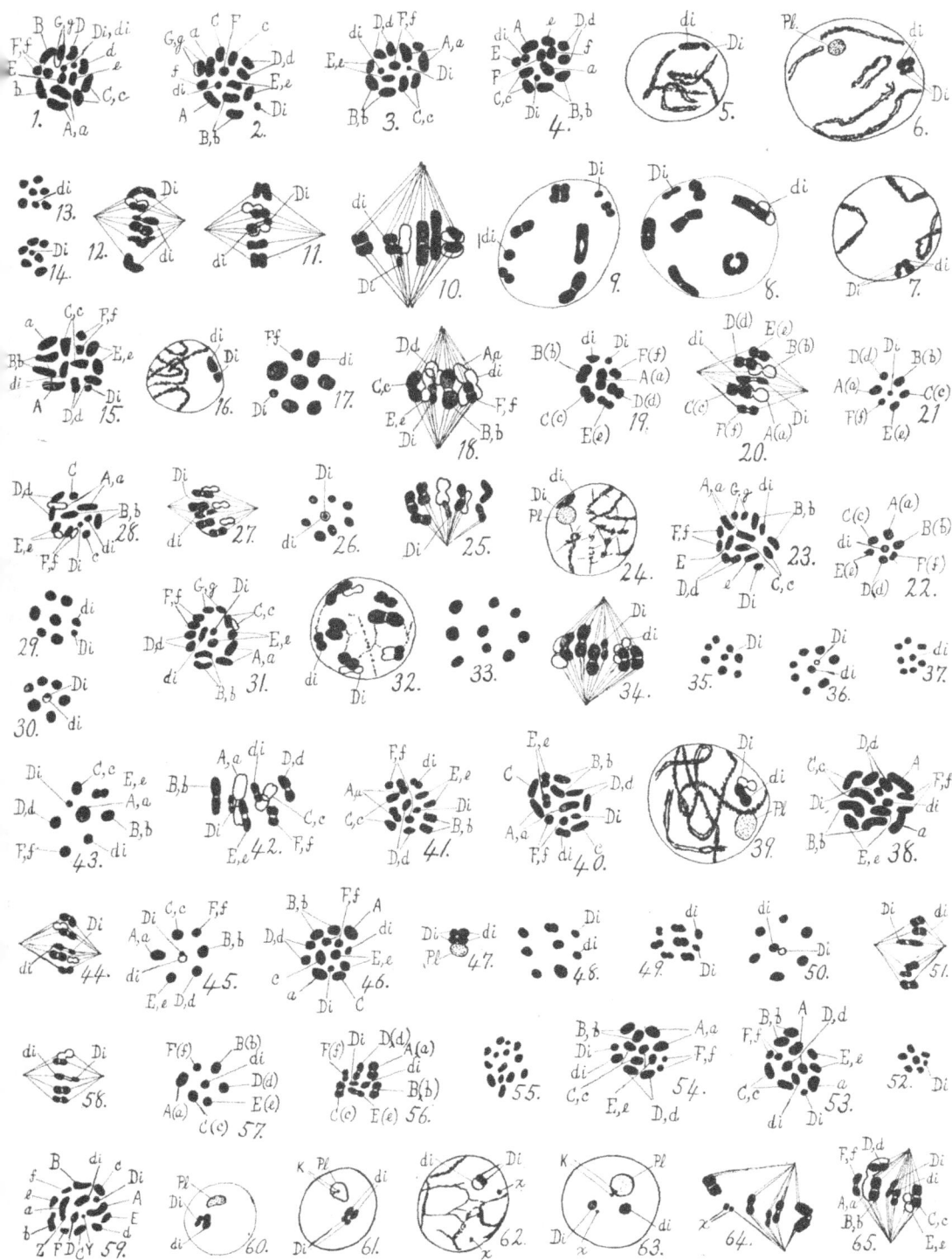

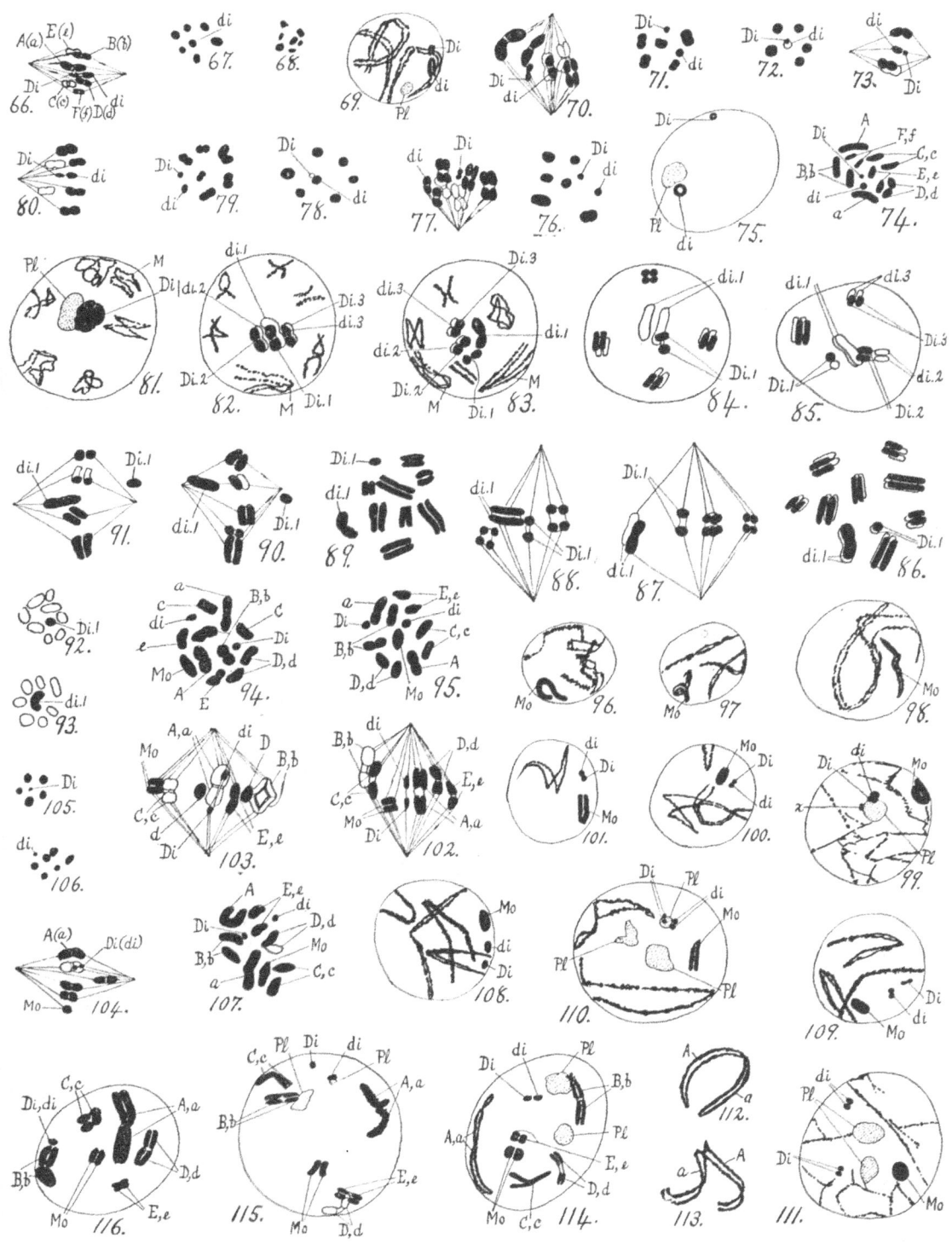

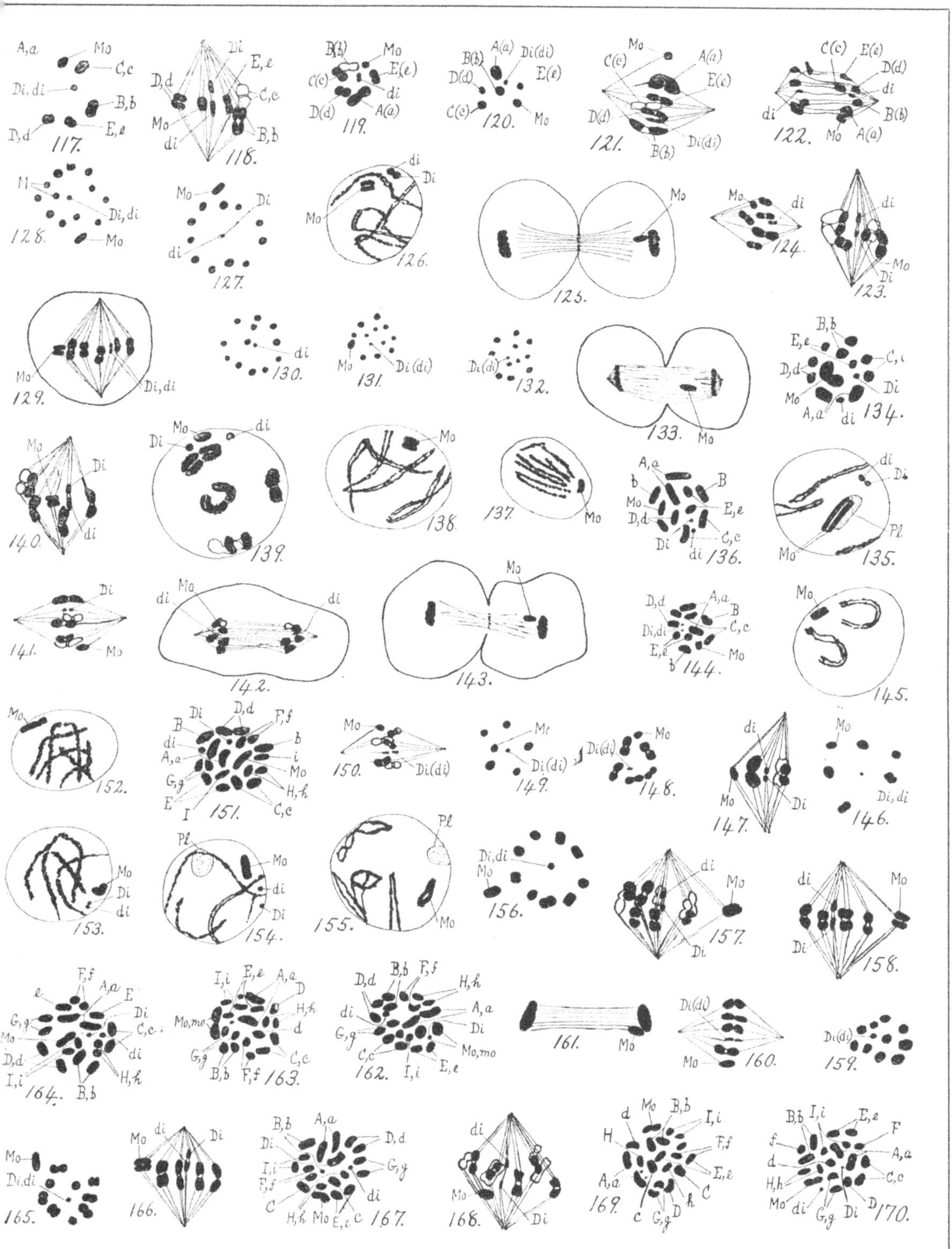

MONTGOMERY.—CHROMOSOMES IN THE SPERMATOGENESIS OF THE HEMIPTERA HETEROPTERA.

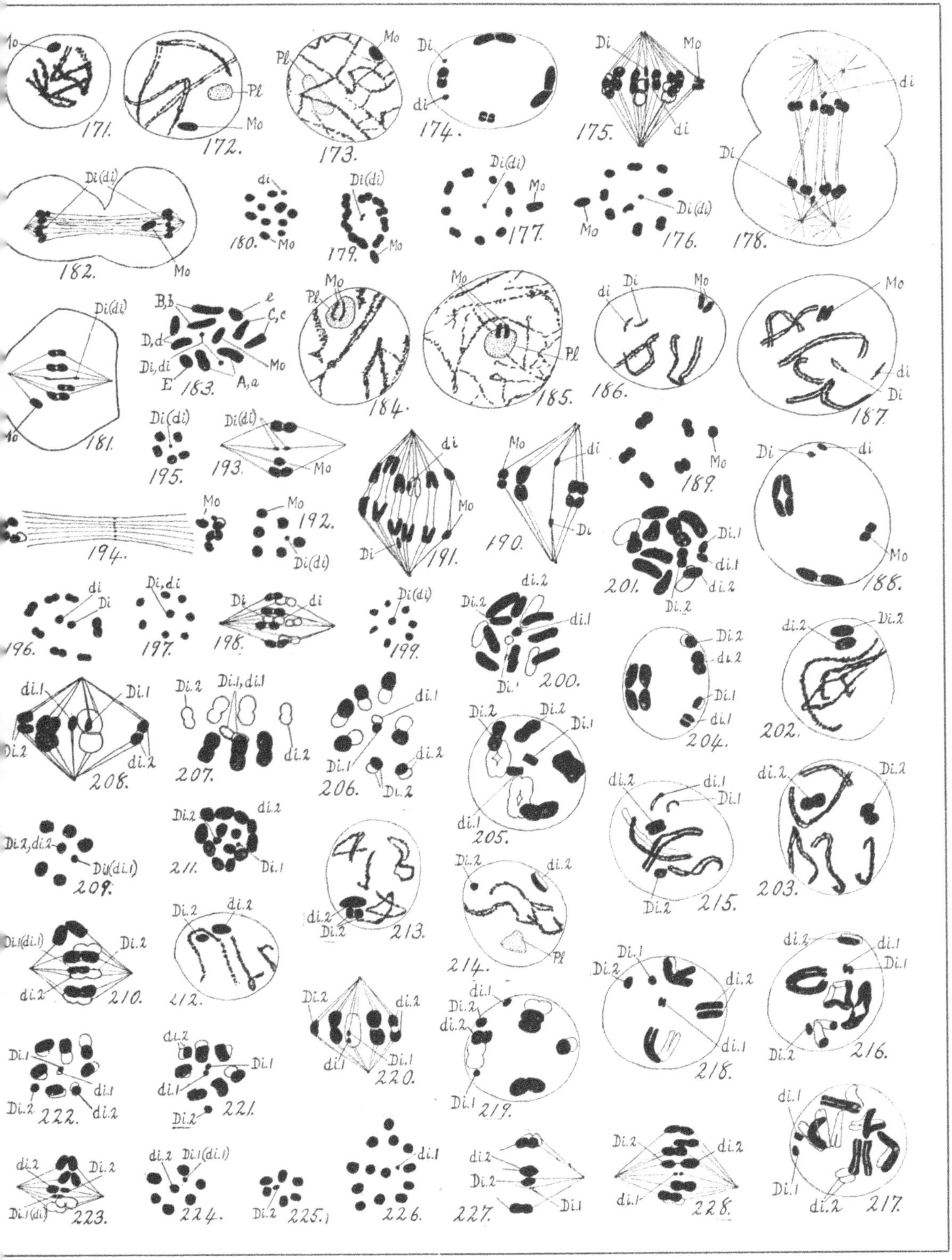

MONTGOMERY.—CHROMOSOMES IN THE SPERMATOGENESIS OF THE HEMIPTERA HETEROPTERA.

PLATE XIII.

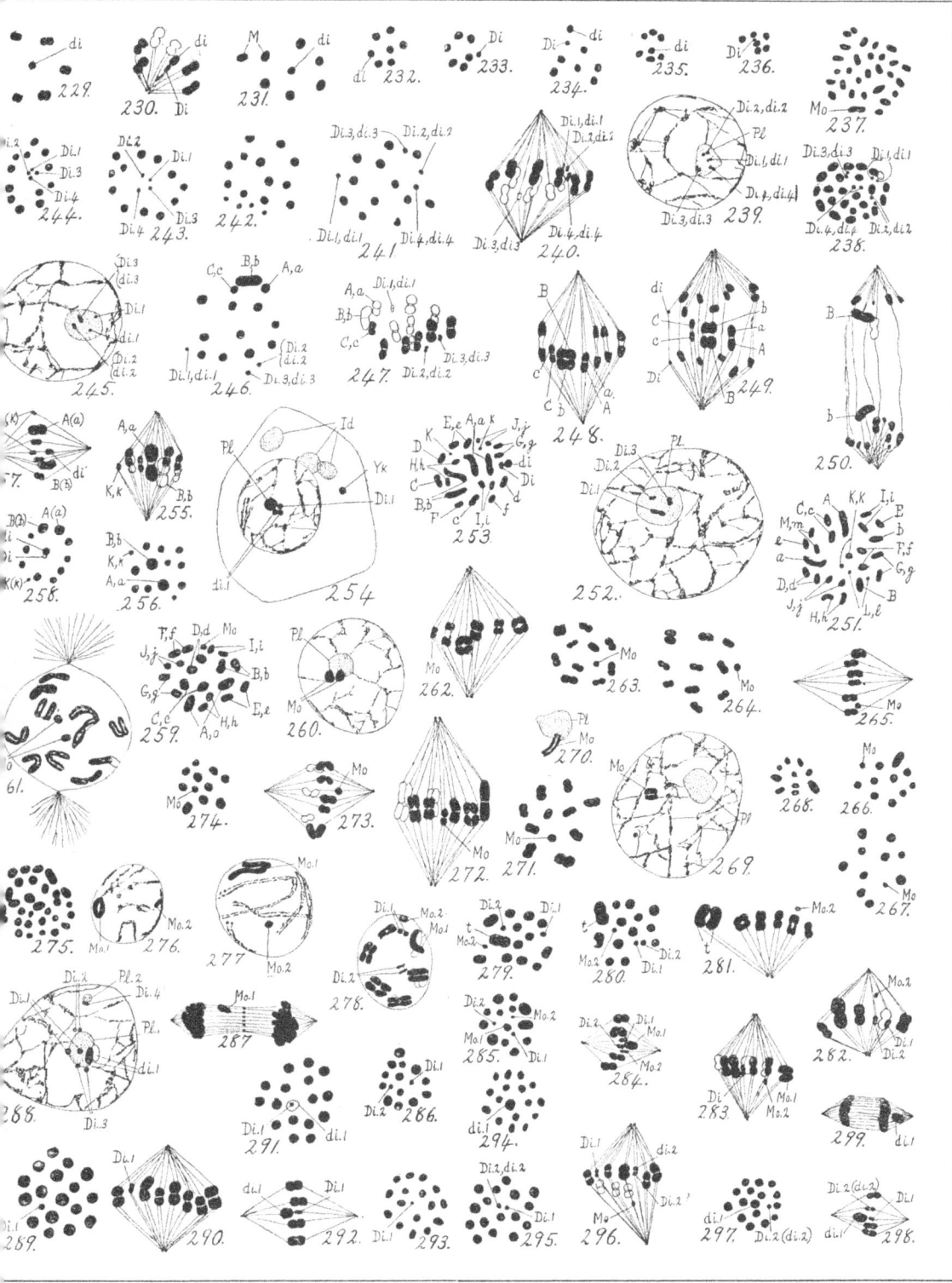

www.ingramcontent.com/pod-product-compliance
Lightning Source LLC
Chambersburg PA
CBHW081335190326
41458CB00018B/6008